What are the challenges involved in protecting biodiversity in tropical terrestrial and coastal ecosystems? What practical lessons can be learned from conservation projects? And what are the procedures and attitudes of governments, NGOs, donor agencies, development banks, and consulting firms? These key questions are all answered, drawing on the author's extensive experience of conservation projects in Malaysia, Nigeria, China, Indonesia, the Philippines, and Costa Rica. Project descriptions are used to illustrate the growth of two important themes in conservation: increasing the awareness of the economic value of biodiversity among decision makers, and enabling and encouraging local people to participate in designing and implementing projects. The first part of the book sets out a series of case histories, while the last five chapters analyse these examples and much other information, setting out conclusions and guidelines to help design conservtion projects which are practical and effective, and which can be duplicated within existing political, administrative and economic systems, yet are also more complete, more inclusive, and more durable than those designed in the recent past.

T0296955

Designing conservation projects

Designing conservation projects

Julian Caldecott ˋ

CAMBRIDGE
UNIVERSITY PRESS

CAMBRIDGE UNIVERSITY PRESS
Cambridge, New York, Melbourne, Madrid, Cape Town, Singapore, São Paulo, Delhi

Cambridge University Press
The Edinburgh Building, Cambridge CB2 8RU, UK

Published in the United States of America by Cambridge University Press, New York

www.cambridge.org
Information on this title: www.cambridge.org/9780521117968

First published 1996
This digitally printed version 2009

A catalogue record for this publication is available from the British Library

Library of Congress Cataloguing in Publication data
Caldecott, Julian Oliver.
Designing conservation projects / by Julian Caldecott.
p.   cm.
Includes bibliographical references (p. 297) and index.
ISBN 0 521 47328 4 (hc)
1. Conservation projects (Natural resources)   2. Biological diversity conservation.   I. Title.
S494.5.P76C35 1996
333.95'16'0913 – dc20   95-46346 CIP

ISBN 978-0-521-47328-6 hardback
ISBN 978-0-521-11796-8 paperback

For Benjamin Lian Caldecott.

Contents

Contents

Foreword

Julian Caldecott has done what many of the conservation veterans should do, but almost none do. He has taken time off to reflect and put his experiences down in what is more than a very readable book. This is a 'must read' book. It is the human ecology of seven tropical attempts to set up nature reserves with sufficient size, social connection, and sustainability to have a serious chance of surving into perpetuity. Federal versus States' rights and demands. Rural dwellers versus central governments. The natural history of poverty and wealth. Dynamics of humans in the harvest of 'wild' organisms from the wildland crop. Dictatorships versus democracies at all scales. Who owns it, me or you? Who is allowed to buy it? Who is allowed to write about it? This is the arena in which tropical conservation occurs, and it is the dynamic of this arena that will determine whether we still have any sustainable remnant of tropical wildland biodiversity 100 years from now.

This book should be read by all mid-level management in natural resources: a look in the mirror. This book should be read by the tropical ruling class: see what tugs at them. This book should be read by the work-a-day folks: see what swirls around them. The stories are told from the viewpoint of a sergeant and lieutenant, in the language of communication rather than as an academic study. This is a book for wildland policy makers from farmer to president: they all make policy, just at different scales. It should be obligatory in the initial orientation of incoming personnel at any level to Peace Corps, World Bank, international development agencies, conservation agencies, consultanting firms, and the like. And we need many more such detailed natural histories of particular projects, told as candidly and boldly as the author feels personally secure in doing. Sadly, they can never be as accurate and detailed as the reader would like. Making public many of the details can seriously damage the ongoing portions of the project and rekindle buried animosities.

Over and over the story emerges that the neighbors generally care more about the fate of conservable wildlands than do those who are more geographically or sociologically distant, except for the occasional (usually) middle- to upper-class outcast, who for a variety of reasons, picks up the lonely wildland conservation flag but is not generally appreciated by his or her (competing) peers. Put bluntly, the decision-making class in a tropical country generally views large blocks of wildlands as part of their income stream, thereby pitting the conservationist directly against national political power.

At the level of the environmental insult, what it all boils down to is failing to treat the wildlands as a crop and a pasture to be planned, invested, researched, designed, sustainably harvested, but rather treating it as a common for a village whose social structure has decayed. Ignorance or administrative chaos? I would say the latter, coupled with a very human 'finders keepers' rule. If conserved wildlands are to survive, wild meat and market have to become just another kind of cattle, and, if carefully husbanded, is no more likely to go extinct than is a farmer's cattle herd and the Siamese cat. Ecotourists are a better kind of cattle. Lead molecules for the pharmaceutical industry are greenhouse flowers – harvest and grow with care and precision, under contract, for a known market. Precision logging is picking apples.

There is no such thing as a donor. Everyone is purchasing something – euphemistically called 'fulfilling an agenda.' The art locally is combining these agendas for their synergisms rather than letting them Balkanize. The art nationally and internationally is giving legitimacy to biodiversity development, to ecosystem development, to recognizing wildlands conserved for their non-damaging use as an interesting, significant, and viable major form of land-use.

All conservation is a form of zoning. People submit to zoning to the degree that they perceive that they gain. And the presence of zoning for any area significantly large enough to be a viable conservation area implies top–down action impacting many economic sectors, many individuals with very different life goals, and various discordant but geographically separate cultural units. Top–down interactions are, however, in whatever conservation budget, expensive. This theme is played out over and over again in Caldecott's analysis of his case studies in the second half of the book.

This book should be read and discussed in detail by any course in conservation biology, developing country politics, contemporary anthropology, and national and local tropical economics. And read slowly. Grasp what those simple sentences really mean. Each case study chapter should be read as a unit, as though it were an autecological study of a single species of animal and its interactions with the social, biological, historical, geological interactions that it has with the others around it.

This book is rich in macrofacts, the sort that need to be grasped to put the nation-level context in place, the context in which real conservation lives or dies. By presenting a detailed natural history of each of seven cases first, and then by generalizing among them, Caldecott presents two very different views of the same processes, two different approaches – one for the person who builds from specifics, and one for the person who dissects from generalities. These generalities are a crude pass at the evolution of what might be called 'sustainable administration and harvest rules for a wildland crop.' Such rules have been evolved, used, discarded over and over throughout human history – the two most extreme sets of rules being 'clear and replace with domesticates' and 'no human presence allowed.' The former is wildland extinction up close, the latter is wildland extinction delayed. Caldecott is aiming for a middle ground, a middle ground that lends itself to few overarching rules, and much energy expended on local and national fine-tuning.

Each case study is still ongoing, each a mixed success, each contains many examples of use to the person who is involved in conservation projects. Read each as a story. They do have the property that they will be most easily grasped by those who have been through this already, but still a careful reading by the entry-level person will impart much. There is no reason that we have to reinvent this wheel. Caldecott has done a superb job of outlining a series of (unfortunately) battles as they played out, and then Monday-morning quarterbacking them. THIS is a conservation biology textbook. There is no way that a person, be it the novice or the veteran, can avoid gaining major tools and having new ideas from reading these accounts and their analyses. This book will be of enormous assistance to the sustainable survival of our wild land biodiversity and the ecosystems that they constitute.

Julian has taken the time out of combat to write it down. You can smell the tension. It seems so simple – describe and bring it to the world's attention. The price has been very high. The world of conservation can pay some of it back by reading and seeing what is here.

Daniel H. Janzen
University of Pennsylvania

Preface

Biodiversity can be conserved through an alliance of local people and other groups, but the context varies since some threats can be solved by communities acting alone, while others are too potent for local people to meet without help. Secure land tenure, for example, can give people the authority to resist encroachment, but well-armed poaching gangs can overwhelm such defences, and so too can planning failures which result in ill-placed roads and logging concessions. Local people often have good reason to help conserve the ecosystems on which they depend, but to do so they must have confidence and power enough to protect their own interests.

This autonomy should be supported by proper central planning and by properly enforced laws, including those which provide for consistent, long-term investment in protecting nature reserves with local consent. Such reserves are valuable in all sorts of ways, but protecting them can be expensive. Hence the need for the greatest possible cost-effectiveness in conservation action (the main subject of this book), and for a proper balance between the powers and responsibilities of different levels of government in undertaking it (the subject of other recent studies[1]

I have tried to give this book as many points of contact with real life as possible, using only observations for which I can vouch personally. Before beginning the case studies, I want to tell a story showing that conservation neeed not be as complex as it sometimes seems. It concerns the fishing vilages (or *barangays*) of Dacu and Oeste, which face each other across a small bay on the island of Libucan in the Samar Sea, in Philippine waters. Inland, there are coconut palms, with patches cleared for temporary farms, or else under tall grass or bamboo. The bay and nearby waters are heavily fished

---

[1] See chapters in *Decentralization and Biodiversity Conservation* (edited by E. Lutz & J. O. Caldecott). The World Bank (Washington, DC, USA).

by local residents and others, including trawlers owned and operated by outsiders.

I visited both villages in July 1995, with a team from the WESAMAR programme. The first stop was at Dacu, where we introduced ourselves, and told the crowd that gathered in the market-place a little about WESAMAR. We explained that it is funded by the European Union, and that it aims to promote sustainable development by meeting community needs identified through open planning. We ended by admiring Libucan's environment, and by saying that local people should be able to teach WESAMAR much about environmental management. The people answered in some amazement, by describing their worries over the collapse of fish stocks in the area.

We asked the group what they had done about this, and they explained that they had passed *barangay* ordinances to ban dynamite-fishing and trawling, but that these were hard to enforce because friendships and family relationships created too many 'special cases'. We asked them how they would solve this new problem, and the consensus idea was to create a fish sanctuary. The aim would be to ban all fishing inside the bay, since this would apply to everyone equally and would be easy to enforce, while allowing the fish stocks to recover. The people were aware, though, that this action would need to be supported by *barangay* Oeste across the bay. We therefore offered to take someone from Dacu over to Oeste in our boat, to see if the people there had the same idea. This offer was accepted, a volunteer was chosen, and we set off after lunch.

The meeting at Oeste quickly showed that people there also saw a problem with fish stocks, and that they would like to see outside fishers excluded from the bay. Dacu's proposal for a fish sanctuary prompted some clapping and much talking. The eventual response was that, rather than a fish sanctuary, the people at Oeste would prefer to allow continued fishing in the bay by local residents, using hook-and-line, gill nets and trawls. There was thus some difference of opinion between Dacu and Oeste. At this point, we said that WESAMAR could provide facilitation services for a joint meeting of the two communities, with the aim of helping them find a common fishery management strategy, and this proposal was welcomed by both sides.

A few hours spent listening to people on Libucan had shown that, without previous guidance by outsiders, they were well aware of the following:

- that a problem existed with an important environmental resource;
- that it was linked to excessive human use of the resource;
- that it could not be solved easily by regulation aimed only at harvesting methods;

- that it might be solved by regulation aimed at ending open-access harvesting;
- that closure of at least part of the area to exploitation would allow regeneration of the resource;
- that collective action with other stakeholders would be needed to make any strategy effective;
- that not all stakeholders were equally dependent on the resource or had equal rights to manage it; and
- that divergent interests among stakeholders needed to be settled by dialogue and compromise.

Thus they understood many key principles of environmental management, on which they could build an effective solution to a challenging and important conservation problem. It is easy to imagine that further dialogue will produce an agreement to exclude outside fishers from the bay, to ban trawling (once both groups understand the damage it does to the benthos, and the consequences of this for the fishery), to continue hook-and-line and gill-net fishing (these having relatively benign effects), and perhaps to set aside part of the bay as a sanctuary. WESAMAR's total investment in beginning this process was the time spent visiting the two villages, and a commitment to facilitate a meeting between them. Whatever the detailed outcome of this and subsequent experiments, it will be owned by the people of Libucan, and is likely to benefit them as well as the marine life under their stewardship.

Acknowledgements

Special thanks are due to my wife, Serena Tan Caldecott, for putting up with me, and for helping with everything since 1980.

The following people commented on one or more chapters of this book, or otherwise helped in its preparation: Nick Ashton-Jones, Richard Barnwell, Sultana Bashir, Liz Bennett, John Bennett, Susannah Bleakley, Stratford Caldecott, Serena Tan Caldecott, Shirley Conover, Alan Crowden, Peter Guy, Liza Gadsby, Roger Hammond, Michael Harrison, Shelagh Heard, Martin Holdgate, Peter Jenkins, Phil Marshall, John Oates, Shane Rosenthal, Gerry Shaper, Lorna Shaper, and Tony Whitten. Although they contributed greatly to the book, they cannot be held responsible for its remaining shortcomings.

The following people all helped in an important way to shape the experience and ideas summarized in this book: Ibu Poppy Abdoelkadir, Ibu Evie Adipati, Jackie Alder, Bpk Hadi S. Alikodra, Peggy Allcott, Richard K. G. Allen, Pius A. Anadu, Q. B. O. (Tunje) Antonio, Nick J. Ashton-Jones, Karen Radel Aylward, Janet Barber, Ed Barbier, Richard Barnwell, David Barton, Eamonn Barrett, Kenny Bell, John G. Bennett, Elizabeth L. Bennett, Russell A. Betts, Claire Billington, Diane Blachford, Susannah Bleakley, Catherine Bickersteth, Paul O. Bisong, Charles Brook, Peter Brosius, Veronika J. Brzeski, Jo Burgess, Datuk Leo Chai, Paul Chai, Barry Coates, Mark Collins, Shirley A. M. Conover, Lisa M. Curran, Chuck L. Darsono, Nassim Datoo, Jean Deong, Gill Dias, John H. Dick, Michael R. Dove, John Ducker, Clement O. Ebin, Clement A. Edet, David Edwards, Margaret Ekerete, Ibu Euis Ekwati, Joanna Elliott, Comfort Emden, Okon Eniang, Paul Erokoro, Felix Esi.

T. F. Fameso, Franklin Farrow, Carlos Fernandez, Virgilio A. Fernandez, Brian Ferrante, H. Benjamin Fisher, Jefferson M. Fox, Marty S. Fujita, Elizabeth (Liza) Gadsby, Helen Gardiner, Rodrigo Gámez Lobo, Gerry Glazier, Hester Godwin, Arnold Golden, Michael Green, Brian Groombridge, Peter

Guy, Frances Hall, Phil Hall, Lawrence S. Hamilton, Roger Hammond, John Hanks, Caroline Harcourt, Tom Harris, Michael J. S. Harrison, Jeremy Harrison, Shelagh C. Heard, David Hitchcock, Martin W. Holdgate, Mike D. Holland, John Howell, Margaret Hoyle, A. Jimmy Ibanga, Maurice Iji, Ibrahim Inahoro, Mark Infield, Uwem Ité, Daniel H. Janzen, Martin Jenkins, Peter D. Jenkins, Timothy C. Jessup, Jorge A. Jiménez, Tim Johnson, Rebekah Kanter, Mike Kavanagh, Georges Kawaja, Val Kapos, Margaret F. Kinnaird, Carolyn Knight, William D. Knight, Thomas Kunzel.

David Labang, Sonia Lagos-Witte, Ruth Laidlaw, Jayl Langub, Pauline Lawrence, Jennifer Leith, Balang Lemulun, A. P. (Tasso) Leventis, Tania Li, Bindu Lohani, P. H. C. (Bing) Lucas, Alistair MacDonald, Kathy MacKinnon, John Makong, Clive Marsh, Peter Martin, John Maturbongs, Jeffrey A. McNeely, Christian (Kit) Milner, David Moles, John Mshelbwala, Justin R. J. H. Mundy, Osé Murang, Steve T. Murphy, Helen Newing, Steven M. Newman, Philip Ngau, Ibu Siti Nissa Mardiah, John F. Oates, Austin Obiekezie, Timothy G. O'Brien, Jonathan C. Okafor, David U. U. Okali, O. Ojong, Herman Ongkiko, Mary C. Pearl, Ronald Petocz, Effren (Goyo) Piczon, Richard W. Pollard, Clifton Potter, Nicki Potter, Robert Prescott-Allen, S. Tahir Qadri, Manuel Ramirez Umaña, Jon C. Reid, Raul Repulda, H. Jack Ruitenbeek, Marcus Robbins.

A. R. K. Saba, Minister Emil Salim, Elvira Sancho, Bpk Didin Sastrapradja, Ibu Setijati D. Sastrapradja, Goetz Schuerholz, Ana Sittenfeld, Norman Sloan, Bpk Soenartono Adisoemarto, Bpk Dr R. E. Soeriaatmadja, Bpk Soetikno Wirjoatmodjo, Bpk Aca Sughandy, Francis Sullivan, Manfred Taege, Jasmine Tan, Johannes Ter Haar, Jane Thornback, Lise Tonelli, Zena Tooze, Beatriz Torres, Paul V. Turner, Danilo M. Valencia, Caroline Vandersluys, Gwen Vaughan, Alban de Villepin, Richard Viner, John Waddel, John C. Ward, Barnabas Watopa, Reginald A. Watson, Dominic White, Judy White, Tony Whitten, Jane Whitten, Clive Wicks, Wendell Wilson, Susan M. Woods, Sejal Worah, Penny Zeng Huifang, Daniel G. Zevin and Richard Zink.

The following agencies provided material support for the work reported here. In Sarawak: WWF Malaysia, Earthlife Foundation, LSB Leakey Foundation, Leverhulme Foundation, Land Associates International Ltd. In Nigeria: WWF UK, Overseas Development Administration, Commission of the European Communities, Fountain Renewable Resources Ltd. In Indonesia: Dalhousie University (EMDI Project), Asian Development Bank, DHV Consultants, WWF Indonesia Programme, LTS International Ltd, Fountain Renewable Resources Ltd. In China: Environmental Resources Management Ltd, Asian Development Bank. In the Philippines: Dalhousie University

(ERMP), Richard Woodroofe & Associates Ltd (WESAMAR Programme), Commission of the European Communities. In Costa Rica: Dalhousie University (EMDI Project), Dalhousie University (ERMP), Living Earth Foundation.

Permission to use copyright material was kindly provided by: Nick Ashton-Jones, Asian Development Bank, Robert Chambers, Dalhousie University, Environmental Resources Management, National Biodiversity Institute of Costa Rica, Ronnie Siakor, WWF Indonesia Programme, WWF UK, Westview Press, and Tony Whitten.

Figure acknowledgements:

Illustrations on pages 5, 31, 62, 84, 111, 135 drawn by Rachel Caldecott-Thornton; p. 5 redrawn from *The Natives of Sarawak and British North Borneo*, volume II by Henry Ling Roth (first published by Truslove & Hanson, London 1896); pp. 31, 111, 135 based on items in the collection of the author; p. 62 redrawn from Whitten (1982).

Fig. 11.1 from Chambers (1983); Fig. 11.2 from Siakor & Ashton-Jones (1992); Fig. 11.3 redrawn by Paula Youens after illustrations in Anonymous (undated).

All maps by Malcolm Barnes based partly on:
Fig. 2.1, Payne, J. F., Francis, C. M., & Phillipps, K. (1985) *A Field Guide to the Mammals of Borneo*, The Sabah Society (Kota Kinabalu, Sabah, Malaysia) and World Wide Fund for Nature (WWF, Kuala Lumpur, Malaysia); Fig. 2.2, Caldecott (1988e); Fig. 3.1, Forest (1993); Figs. 3.2 & 3.4, Caldecott, Oates, & Ruitenbeek (1990); Fig. 3.3, Caldecott, Bennett, & Ruitenbeek (1989); Fig. 4.2, Whitten (1982); Figs. 4.3, 5.5, & 7.1, Collins, Sayer, & Whitmore (1991); Fig. 5.2, ADB (1992c); Figs. 5.3 & 5.4, ADB (1992d); Fig. 5.6, Caldecott (1993a); Figs. 6.2 & 6.3, National Biodiversity Institute of Costa Rica; Fig. 7.2, Irian Jaya Provincial Government planning map of 1987; Fig. 7.3, WWF Indonesia Programme.

1

Introduction

Thousands of the wild species which were alive in the 1950s and 1960s no longer exist. We cannot bring them back, but we can learn from their deaths. While debate continues on how to make human development more sustainable, an immediate task is to protect the diversity that remains against known hazards. These hazards are potent, and ideas on how to resist them are still evolving. There is an urgent need for measures which are practical and effective, and which can be duplicated within existing political, administrative, and economic systems. This book advises on the design of conservation projects which meet these criteria, yet are more complete, more inclusive and more durable than those designed in the recent past.

The book is mainly about conserving wild species, and the life styles and cultures which depend upon them. Most of these species occur in the tropics, and cannot survive outside the natural ecosystems to which they are adapted. Very few can be kept alive for long or at reasonable cost in captivity or in artificial habitats. The book therefore emphasizes conserving natural ecosystems in the tropics, especially those that occur in areas where it seems feasible to protect them. These areas are called nature reserves, and are defined as places containing viable samples of natural ecosystems which are, or might be, or should be, set aside by law or custom mainly for conservation. A conservation project, then, is a planned undertaking which aims to protect a nature reserve.

Nature reserves would not exist and would not be necessary if the short-term needs of people did not conflict with those of wild species and natural ecosystems. But such conflicts do exist, and the main task of conservation projects is to resolve them. To do this for any threatened nature reserve, a project must have a wider perspective than that of the reserve itself, and a longer-term perspective than that of the people who live nearby, or those who might wish to exploit it and are careless of damaging it. This is why

1

the designers of modern conservation projects try to reach out beyond reserve boundaries, seeking to solve problems within the larger ecological and economic regions around them (e.g. Wells, 1989; Brandon & Wells, 1992; Wells, Brandon & Hannah, 1992; Wells & Brandon, 1993).

When conservationists began reaching out in this way in the early 1980s, they found the need to learn more about people, about how communities work and how decisions are made. The ideas of anthropology, education, politics and economics had to be absorbed, and this led to the creation of the new, synthetic discipline of conservation. This discipline is now able to propose feasible alternatives to the destruction of biodiversity, in ways which can be appreciated by powerful decision makers as well as by the public in any country. The science of biology meanwhile, where many conservationists began their careers, has become far better able to provide technical guidance to those seeking to preserve biodiversity.

One of the aims of this book is to describe and illustrate the journey which conservationists made collectively between the early 1980s and the mid-1990s. This involved many different events, and every individual experienced different species and ecosystems, stakeholders and interest groups, among them local people, local, national, and international NGOs, development agencies, and governments. The critical period was punctuated by the great, consensus-building World Congresses on National Parks, in Bali, Indonesia in 1982, and Caracas, Venezuela in 1992. Other landmarks included the review publications that emerged from those congresses (e.g. MacKinnon *et al.*, 1986; Kemf, 1993; Barzetti, 1993).

The story is told here through a series of field projects, starting in Sarawak in 1984–8, and then in Nigeria in 1988–90, China in 1991, western Indonesia in 1992, Costa Rica in 1992–3, the Philippines in 1993 and Irian Jaya in 1994. It therefore begins in one immense rain forest in the interior of Borneo, and ends in another in the interior of New Guinea. It describes some of the species and people in a wide variety of tropical environments, the effects upon them of exploitation and temporary development, and some efforts to avoid or resolve the conflicts that arose. There are two main themes. The first is about raising awareness of the economic value of wild species and natural ecosystems, in the hope that decision makers will then treat them better. The second concerns helping local people achieve conservation themselves, through direct involvement in projects and in resource management.

All of the project descriptions are based on the personal experience of the author. They provide real-life examples of most of the issues and techniques which are discussed in the final five chapters of the book, where empirical observations are meant to blend with theory. These chapters aim to show that there are no easy, universal answers in conservation, and that all guide-lines must be applied carefully according to local circumstances. The chal-lenge to the reader is to accept that nowhere is the same as anywhere else and that places can change quickly with human influence, and, having accepted it, to design projects which are locally appropriate. The final aim of this book is to show that this is possible, and how it can be done.

2

Baram River, Sarawak

2.1 Conservation issues

2.1.1 Land, people, and biodiversity

Borneo is about 740 000 km² in area, and is divided into seven political units: the Malaysian States of Sabah and Sarawak, the Sultanate of Brunei, and the four Indonesian Provinces of East, West, Central and South Kalimantan (Fig. 2.1). At about 124 450 km², Sarawak is the largest of the thirteen states of Malaysia, which it joined in 1963 after nearly 100 years as an independent country under the Brooke Rajahs, and brief periods of Japanese occupation and British colonial rule. This history sets Sarawak apart from other countries in South-east Asia, all of which other than Thailand had been ruled directly by powers from outside the region.

The Brooke administration in Sarawak had an unusually consultative style of government, which was summed up by Rajah Charles Brooke as follows: 'A government such as that of Sarawak may start from things as we find them, putting its veto on what is dangerous or unjust, and supporting what is fair and equitable in the usages of the natives, and letting systems of legislation wait upon occasion. When new wants are felt, it examines and provides for them by measures rather made on the spot than imported from abroad; and to ensure that these shall not be contrary to native customs, the consent of the people is gained for them before they are put into force' (the *Sarawak Gazette* of 1872, quoted in Brooke, 1913: xxiii–xxiv). This approach established a persistent tradition of tolerance and intimacy between government and rural people, which continued into the 1980s.

Sarawak's population was about 1.6 million in 1989, of which almost 30% were Iban, 29% were Chinese, 20% were Malay, 9% were Bidayuh, 6% were Melanau, and the rest belonged either to the diverse Orang Ulu group (meaning *people of the headwaters*, about 5% of the total, comprising Kayan,

4

Shield design by Kenyah people of the upper Baram region of Sarawak.

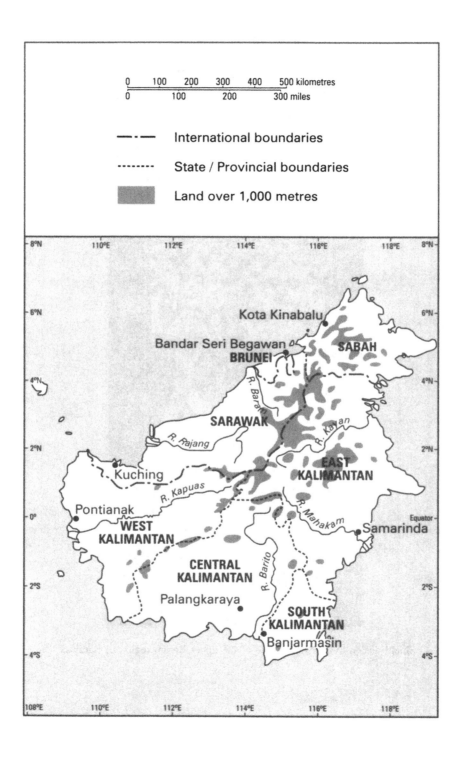

0	100	200	300	400	500 kilometres	
0		100		200		300 miles

—·—·— International boundaries

········· State / Provincial boundaries

▩ Land over 1,000 metres

Kota Kinabalu

Bandar Seri Begawan
BRUNEI

SABAH

R. Barah

SARAWAK

R. Rajang

R. Kayan

EAST
KALIMANTAN

Kuching

R. Kapuas

Pontianak

R. Mahakam

Equator
Samarinda

WEST
KALIMANTAN

CENTRAL
KALIMANTAN

R. Barito

Palangkaraya

SOUTH
KALIMANTAN

Banjarmasin

Kenyah, Kelabit, Lun Bawang and other peoples), to other groups (about 1%), or to the small population of Penans. Some Penans still followed a semi-nomadic, forest-dwelling, hunter-gathering life style (Brosius, 1986). The Malays and Chinese tended to live in urban and coastal areas, while the Iban, Bidayuh, and Orang Ulu were mainly inland farmers, characteristically cultivating dryland rice, and living in longhouses. The latter are single buildings housing whole villages, and are a famous feature of Bornean architecture.

Like the rest of Borneo, Sarawak has a moist equatorial climate, and most places have an average annual rainfall of 3000–5000 mm. In the lowlands and coastal plains there are about 1740 km² of mangrove forest growing in tidally flooded muddy areas, and about 14 700 km² of peat swamp forest growing at or near sea level, often on deep peat deposits. About 10% of the State comprises rolling and moderately steep, low hills, and 70% is steep hilly and mountainous country (WWF, 1985). The natural vegetation type which dominates this rugged interior is tropical evergreen rain forest (Whitmore, 1984), which changes with altitude from lowland mixed diptero-carp forest (below about 700 m), to hill dipterocarp forest (from about 700–1200 m), to lower montane rain forest (from about 1200–1500 m), and eventually to upper montane rain forest (above about 1500 m). Significant areas of low-stature heath forest occur throughout the interior, reflecting the distribution of extremely nutrient-poor and often white-sand soils.

The lowland dipterocarp forests are the most species-rich of these forest types, and up to 2300 species of tree have been recorded in the northern lowlands of Sarawak, compared with less than 250 found in peat-swamp forests and about 850 in heath forests (Whitmore, 1984). Species richness declines with elevation, although endemism may increase and several hills in the northern Sarawak area comprise centres of plant diversity (WCMC, 1992; WWF & IUCN, 1995). The same is true among birds, for example on Gunung Mulu in northern Sarawak, where 171 species in lowland forest are reduced to only 12 at 1300 m (Whitmore, 1984); the mountains of northern Borneo comprise an endemic bird area to which 26 restricted-range birds are confined (ICBP, 1992).

Fig. 2.1 The island of Borneo

The lowland and hill 'dipterocarp' forests are so called because many of their large trees belong to the family Dipterocarpaceae. The abundance of these trees is a common feature of lowland and hill forests in Borneo, Java, Sumatra, and the Malay Peninsula. Such ecological similarities among these land masses are due to the fact that they all lie on the same continental shelf of South-east Asia, the Sunda Shelf, and they were linked by dry land at times of low sea level over the last 2 million years (Chapter 4). The ecological influence of dipterocarp trees is due to their abundance and their fruiting behaviour, which tends to be synchronized within and among species. It is also irregular, resulting in massive fruit-crops (mastings) at unpredictable 2–5 year intervals (Janzen, 1974).

This mast-fruiting is thought to be an adaptation to avoid seed predation, since populations of seed-eating animals are overwhelmed with food during mast years, but starved at other times. This keeps their standing abundance low and limits the damage they can inflict on any one seed crop. Water stress during occasional droughts is believed to provide the main stimulus for masting by dipterocarps. Since many other tree taxa use the same environmental cue to prompt flowering, there is a tendency in dipterocarp forests for the foods available to fruit-eating and seed-eating animals to be either super-abundant or almost absent at any given time. This helps to explain why such animals are rare collectively in dipterocarp forests compared with other rain forests, and also why the biomass of fruit-eating primates is inversely related to dipterocarp abundance within forests that are otherwise similar (Caldecott, 1980; 1986; 1991a; Marsh & Wilson, 1981; Bennett & Caldecott, 1989).

The pattern of fruit and seed availability in dipterocarp forests favours adaptations of high mobility and/or rapid reproduction among animals which depend on such foods. The first allows them to track fruiting activity over wide areas, while the second allows their populations to respond swiftly to unpredictable food supply. Bearded pigs (*Sus barbatus*) show both adaptations. They can travel long distances when tracking food supplies around the Bornean interior, and their populations can erupt lemming-like when they have continuing access to abundant food, especially to dipterocarp seeds (Pfeffer, 1959; Caldecott & Caldecott, 1985; Pfeffer & Caldecott, 1986; Caldecott, 1988a, 1991a, 1991b, 1991c; Caldecott, Blouch, & Macdonald, 1993; Box 2.1). This is the most-hunted species in Borneo, and bearded pigs provide much of the protein and fat eaten by rural people. An important conservation issue arises because dipterocarp timber is a mainstay of the timber industry, and logging thus affects bearded pigs and the people who hunt them. The impact of logging is made greater because dipterocarp seeds and resins are also used by rural people for subsistence and sale.

Box 2.1 Ecology of bearded pigs (from Caldecott, 1991a, 1991b)

Bearded pigs consume roots, fungi, small animals, turtle eggs, carrion, and items from at least 50 genera of plants. Fruit supply controls growth, fattening, and breeding, and the oil-rich seeds of dipterocarps, oaks, and chestnuts are especially important. Highland oak forests produce fruit regularly, and are important as predictable food sources, while dipterocarp forests provide large supplies of food only irregularly by masting. Dipterocarp fruiting is well documented in Sarawak because the seeds of many of these trees are exported commercially, and records have been kept for many decades. In the years since 1899, virtually no dipterocarp seeds were exported in a third of the years, small amounts in most other years, and more than 1000 tonnes in some years. These heavy fruiting episodes were usually isolated and occurred only every 3–5 years. The only exceptions to this pattern in the period 1945–88 were pairs of years with heavy dipterocarp fruiting in 1953–4, 1958–9, 1982–3 and 1986–7. In the 1980s, these events were correlated with repeated droughts linked to El Niño/Southern Oscillation phenomena. Bearded pig numbers erupted in 1954, 1959, 1983, and 1987, but in no other year in this period, suggesting that explosive population growth occurred in response to sustained availability of dipterocarp seeds in consecutive years.

Features which may allow this response include: large average litter size (3–12, depending on mother size); short gestation length (90–120 days) with up to two litters per year; variable but potentially early age at first rut and pregnancy (10–20 months); efficient conversion of dietary fat to body fat; variable but potentially high growth rates; very flexible group sizes (tens to hundreds); synchronization of birth peak with fruit fall (and of the rut with late flowering and flower fall, called the 'confetti effect' because the pigs may use fallen flowers as a visual cue); and travel-adapted features, such as long legs and swimming ability. During the bearded pig eruptions of the 1950s, Pfeffer (1959) described annual population movements in Kalimantan over distances of 250–650 km and, in the 1980s, Caldecott (1988a) tracked herds moving through the upper Baram area of Sarawak at a rate of 8–22 km/month sustained over at least 4–8 months.

Large-scale movements by bearded pig populations have also been reported from the Malay Peninsula (Allen, 1948; Kempe, 1948; Hislop, 1949, 1952, 1955). During these, the pigs were described as moving consistently in one direction, in scattered or condensed herds, over a broad or narrow front, and over a period of several days, weeks, or months. The animals were variously described as being in good, poor, or very poor physical condition, sometimes accompanied by piglets and sometimes not, and regularly swimming across rivers, sometimes coastal bays, and even out to sea. In some cases, the population was thought to retrace its route later, or to follow a circular course to return whence it originally came. In the Malay Peninsula, there is historical evidence that bearded pigs migrated regularly to take advantage of predictable fruiting in camphorwood (*Dryobalanops aromatica*) forests, which have since been felled for their valuable timber.

2.1.2 Shifting cultivation and hunting

The soils of the interior of Sarawak are mostly derived from sedimentary rock and are fragile, deeply leached by the high rainfall, and have low fertility. Since steep terrain is widespread, good agricultural land is very scarce. The low population density of Sarawak as a whole reflects these inhospitable conditions, but three-quarters of this population is concentrated in the western third of the State, away from the interior where human settlements are extremely sparse.

The most widespread form of agriculture in Sarawak is shifting cultivation (also known as swidden or *ladang* farming). This is a general name for any form of agriculture in which fields are cleared by hand and fire, and cropped discontinuously. In established systems, short periods of cropping alternate with longer periods of fallow prior to re-use of the same land, while in pioneer systems, significant amounts of climax vegetation are cleared each year. Large numbers of Iban pioneer shifting cultivators entered Sarawak during the sixteenth Century, arriving from the Kapuas River to the south, and pushed northwards again in a second migration during the nineteenth Century (Freeman, 1970). In doing so they bypassed the more sedentary Bidayuh population of West Sarawak, but came increasingly into conflict with Orang Ulu groups, and also with the Brooke administration (Cramb & Wills, 1990).

By about 1980, some 30 000 km² of Sarawak had been affected by shifting cultivation and about 1000 km² was being cultivated annually (Hatch, 1982). The area of virgin forest cleared each year was uncertain, although the Forest Department estimated it to be about 350 km² at that time. Most shifting cultivation in Sarawak was no longer in the pioneer phase by the early 1980s, although the Iban in particular retained a strong motivation to clear new land, which is traditionally regarded as the source of 'those bountiful and heavy harvests which are the foremost aim of all Iban endeavour' (Freeman, 1970: 130).

Shifting cultivation had been particularly intense in areas below 500 m in elevation, and about 44% of the land area in the western third of the State had been affected by it compared with an average of only about 18% elsewhere. Rural areas in west Sarawak thus typically comprised patchworks of cleared land and regenerating vegetation with little intact forest. Elsewhere, such patchworks followed river valleys, which were often separated by large expanses of forest showing no sign of disturbance. This was the chief difference between the Iban-dominated areas of southern-interior Sarawak, and the homelands of the Orang Ulu in the northern interior, including most of the middle and upper Baram river system.

Rain forests are marginal habitats for people, and to survive there the Penans of Sarawak, who have no traditional agriculture, use as a staple the starch of hill sago palms (*Eugeissona utilis*). Penans are therefore absent in areas where these palms are missing, such as in Sabah. Where such special wild food sources are not available, people cannot occupy rain forests without having access either to trade routes (as among the Mbuti of central Africa; Sayer, Harcourt, & Collins, 1992), or else by practising some form of shifting cultivation to provide staple foods. It can equally well be said, however, that shifting cultivation is not viable without access to rain forest products, since all communities living in or near rain forests derive much of their food supply (and many other useful materials) by hunting, fishing, and gathering in the forest near their homes.

The strong link between shifting cultivation and hunting in Sarawak has long been known (King, 1978), and was investigated further in 1984–6 (Caldecott, 1988a). The latter study was commissioned by the World Wide Fund for Nature (WWF) and the Sarawak Forest Department, because of concern that hunting pressures would damage wildlife populations in the absence of new wildlife management policies and legislation. The study set the scene for later work on the economic role of non-timber forest products, and for the proposed Baram project which is described below.

At the beginning of the hunting study, it was already clear that the interior of Sarawak was very heterogeneous, and that hunting patterns would vary greatly from place to place. It was also clear that bearded pig populations were very mobile, and used areas of hundreds and perhaps thousands of km². The aim of the study was to obtain a quantitative picture of the scale of hunting and the use of wild meat for food throughout the interior of Sarawak, while also documenting the ecology of the main prey animals. Three factors were of great help, the first of which was that the people of the interior of Sarawak were immensely knowledgeable about wildlife and hunting. As Corner (1961: 12) had observed: 'Don't we need the natural history of wild pigs in tropical forests? And who could inform better than the forest-folk of Borneo?' Secondly, they were completely willing to share their observations and opinions about wildlife and also their hunting experiences, perhaps partly because at that time few prey species were protected by law. Lastly, a very large number of useful data had been collected by teachers at rural boarding schools (as records of school meals), by local government officials (as records of trade in shotgun ammunition and dipterocarp seeds), and by traders in wild meat on the main rivers.

The approach used was similar to rapid rural appraisal, in that phenomena were documented quickly over large areas using multiple sources of data (Carruthers & Chambers, 1981; McCracken, Pretty, & Conway, 1988). The main study area covered at first some 8000 km² of the upper Baram, but was later extended to other areas. Rural people were interviewed collectively at 220 longhouses, taking advantage of the architecture, which made it easy to gather people together for meetings. People were also questioned individually concerning nearly 1000 hunting excursions representing a total of more than two person-years of hunting effort. The results were integrated with data on over three longhouse-years of hunting activity collected by informants at three longhouses. They were further compared with an analysis of 2 million daily rations of food at 63 rural boarding-schools during 1984–5, which recorded the consumption of about 203 t of wild and domestic meat. These data were then related to trade figures and other miscellaneous information to present an account of hunting in the interior of the State.

It was found that the methods of hunting used in any given area depended on factors such as local tradition, familiarity with techniques, access to ammunition and lights, and access to markets for meat and trophies, all of which also affected the choice of species taken and the scale of the harvest. Overall, most hunting was by trapping, spearing, blowpiping, and especially by shooting, with nearly two-thirds of all hunted animals dying by gunfire. Among nearly 5500 families surveyed in the interior, there were on average about two hunting dogs and two spears per family, one shotgun per two families, and one blowpipe among four. At the time of the study there were about 60 000 legal shotguns in Sarawak, and an average of nearly 2 million cartridges were being imported each year.

Almost any animal giving a mouthful of meat or more was liable to be captured and eaten in Sarawak, and a variety of products other than meat were used and traded widely. These ranged from ingredients in East Asian traditional medicines, such as pangolin scales and *bezoar* stones (tannin-rich concretions from the guts of porcupines and leaf-monkeys), to the hornbill feathers used for decoration and dancing throughout Borneo. Most data concerned wildlife as food, however, and showed that 80–90% of the wildlife harvest was contributed by sambar deer (*Cervus unicolor*), barking deer (*Muntiacus muntjak*), mouse deer or chevrotains (*Tragulus napu* and *T. javanicus*) and by bearded pigs, with the last being by far the most important prey species. The role of wild meat in nutrition was revealed by school meal records, which showed that, of all meat and fish consumed, bearded pig meat contributed 32%, other wild meat about 7%, fish about 18%, and domestic pork, beef, and chicken 13–16% each. The communities concerned all used rice and other farmed crops as their primary staples, but took more than half of the animal matter they consumed from the wild.

Using different sets of data, three estimates were made of the total weight of wild meat harvested in Sarawak, ranging from 9400 to 26 500 t each year, and the midpoint of 18 000 t was taken as the working figure for other calculations. Its economic value could be estimated in several ways, one being to use government figures on investment in livestock and fishpond development, and the resulting rates of production. To replace Sarawak's wild meat harvest by these means would have cost about US$50 million, although this would have been hard to do in practice. It was argued that the rural population's dependence on wild meat would be an important factor in Sarawak for the foreseeable future, and a strategy was suggested to combine sustainable management of the wildlife harvest with cost-effective investment in protein production at the longhouse level.

The midpoint of 18 000 t worked out to an average consumption of about 12 kg per person per year, although rates of about 54 kg per person per year were recorded for remote, undisturbed areas. These figures were consistent with other data from the interior of Borneo, for example the 320 wild pigs killed at the small Kenyah village of Long Ampung in Kalimantan in 1980 (Colfer *et al.*, in press), and the 49 killed by 16 families in five months of 1977 at the Kenyah village of Long Selatong in Sarawak (Chin, 1985). They were also consistent with rates of wild meat consumption in sub-Saharan African countries, which range up to about 75 kg per person per year from an average of 2–10 kg (Sale, 1981; Prescott-Allen & Prescott-Allen, 1982), and those in Peruvian Amazonia, with a range of 19–168 kg per person per year (Dourojeanni, 1985).

In 1984, the Sarawak State Legislative Assembly began an inquiry into the depletion of Sarawak's flora and fauna (Kavanagh, Abdul Rahim, & Hails, 1989). This gathered evidence from expert witnesses and the general public, and provided an opportunity to bring new information on the scale of hunting in Sarawak to official attention. The inquiry continued until 1988, when new legislation was prepared to update the National Parks Ordinance of 1956 and replace the Wild Life Protection Ordinance of 1958. These new laws were passed in 1991, when the National Parks (Amendment) Ordinance made it possible to gazette small Nature Reserves to protect salt licks and other sites of special importance to wildlife, as well as using the existing options of National Parks and Wildlife Sanctuaries. The Wild Life Protection Ordinance of the same year removed the problem of dual status forest land, and established that Wildlife Sanctuaries could no longer be logged or subjected to silvicultural treatment. It also greatly extended the protection afforded to wild animal and plant species, and made it possible for wildlife management rules to be tightened quickly in response to new threats (E. L. Bennett, personal communication, May 1994).

Table 2.1 *Production and export of logs in Malaysia and Sarawak (from Kavanagh, Abdul Rahim, & Hails, 1989; MASKAYU, 1994)*

Year	Production of logs in		Export of logs from Malaysia (million m³)
	Sarawak (million m³)	Malaysia (million m³)	
1977	4.9	–	–
1978	5.9	–	–
1979	7.6	–	–
1980	8.4	27.9	15.1
1981	8.4	30.7	15.9
1982	11.2	32.7	19.3
1983	10.6	32.8	18.8
1984	11.4	31.1	16.0
1985	12.3	28.7	19.8
1986	11.4	29.9	19.0
1987	13.6	35.1	22.9
1988	–	39.0	20.6
1989	–	41.0	21.1
1990	15.8	39.7	20.4
1991	15.8	39.1	19.3
1992 (est.)	15.3	–	18.2

2.1.3 Logging and people

Sarawak has been a major exporter of timber since the 1960s (Kavanagh, Abdul Rahim, & Hails, 1989; Nectoux & Kuroda, 1989; ITTO, 1990; World Bank, 1991a). The history of logging in Sarawak has been affected by the fact that revenues from off-shore oil exploitation are exclusively Federal, leaving the State Government with timber as its main autonomous source of large-scale finance (Gillis, 1988a). Logging began in the peat swamp forests of the coastal plain, and in the 1970s began to move into the hilly interior which, by the late 1980s, had come to dominate log production. Total production increased steadily from about 5 million m³ yr^{-1} in the late 1970s to about 14 million m³ yr^{-1} in the late 1980s, with higher figures recorded since (Table 2.1). By 1987, the forestry sector provided the jobs of a third of Sarawak's paid workforce, and more than half of the State Government's total revenue.

The amount of timber which might be produced sustainably from Sarawak's forests is uncertain, and depends on assumptions of areas to be used, timing of harvesting, and regulation of logging (ITTO, 1990; World Bank,

1991a). An enquiry by the International Tropical Timber Organization (ITTO, 1990) led to a consensus that logging rates above 10 million $m^3 yr^{-1}$ could not continue for long, and in response Sarawak agreed to limit timber production to this level (Primack, 1991). Otherwise, sustainable logging levels were thought to be from 4 to 9 million $m^3 yr^{-1}$. An annual harvest of 13 million m^3 would mean the logging of all primary forests in Sarawak at least once by the turn of the century (ITTO, 1990).

By the late 1980s, the official target was to retain 50% of Sarawak under permanent forest cover, and about 37% of the land area had been gazetted into the permanent forest estate (PFE). The PFE comprises various kinds of production forest (Protected Forests, Forest Reserves and Communal Forests) and protection forests (National Parks, Wildlife Sanctuaries and later Nature Reserves). It was created partly from lands which much of the rural population viewed as being available for shifting cultivation, or already subject to claim under customary law. About 3% of the PFE was affected by farming (Collins, Sayer, & Whitmore, 1991), and ITTO (1990) estimated that about 5000 km^2 of Sarawak's forests might be affected by claims based on traditional land rights.

When most of the PFE was gazetted during the 1970s and 1980s, therefore, there was an increasing conflict of interest and perception between rural farmers on the one hand, and the Forest Department and timber industry on the other. Each group tended to see intact forest land as a different kind of resource: one as a potential source of rice to be obtained through farming, the other as a potential source of timber to be obtained through logging.

Increasing competition for forest land between farmers and foresters was only one of several related issues which arose in Sarawak during the 1980s. Another concerned the side-effects of logging, which were perceived as damaging to the life style of shifting cultivators. The underlying point is that shifting cultivation requires access not only to land, but also to a wide range of materials for use or sale which can be gathered from surrounding forests and rivers. In Sarawak, these include wild meat and fish, rattan canes, dipterocarp seeds and resin, medicinal plants, and, in case of crop failure, emergency starch sources such as hill sago palms.

Sarawak forestry law recognized the need of shifting cultivators to hunt, fish, and gather non-timber forest products, even within the PFE, where rights to do so could be gazetted in favour of specified communities. There remained the problem that, even where rights were granted to hunt and gather in logged-over portions of the PFE, logging itself may have damaged the resources to which farmers required access. Unless adequately compensated, this damage would be expected to cause the economic position of certain

Table 2.2 *Estimated killing-rates for major human prey species in Sarawak in the mid-1980s, in relation to time since exposure to logging, based on community interviews (from Caldecott, 1988a)*

	Median number of kills/10 families/year:			
	In unlogged	In logged areas – years since first logging:		
Type of animal	areas	1–10	11–20	21–30
Bearded pigs	100	32	14	3
Barking deer	7	1	< 0.5	< 0.5
Mouse deer	7	9	2	2
Porcupines	6	1	1	1
Civets	4	5	2	2
Sambar deer	3	1	1	1
Number of longhouses	24–64	16–41	10–14	7–9
Weight of meat consumed (kg)	3806	1240	534	155
Meat ration (kg/ person/year)	54.4	17.9	7.7	2.2

communities to deteriorate. Logging removes adult dipterocarp trees, and thus reduces the supply of dipterocarp seeds and resin. Mechanized logging at high intensities is also likely to damage rattan stocks, fruit trees, and, through muddying and diesel pollution, fish populations in the rivers. Extensive damage to the forest would also be expected to change its ability to support wildlife populations, for example by reducing food available for bearded pigs, regardless of other factors that may be operating.

Many local informants in the upper Baram were concerned that logging would cause pollution of waterways, loss of fish and game and other benefits, as well as making the environment less attractive and less pleasant to live in. There was some evidence that residents in areas affected by logging and other forms of widespread disturbance did experience a sharp decline in wild meat harvests (Table 2.2). The average ration, estimated from reported harvesting rates, was found to fall from 54 to about 18 kg per person per year during the first decade after logging.

A number of factors in addition to logging were at work here. Logged areas were often exposed to an influx of new hunters, with better transport links for importing hunting equipment (e.g. ammunition, batteries and carbide for lights, and wire snares) and for exporting animal carcasses for sale elsewhere. Different wild species also declined for different reasons, with bearded pigs apparently being vulnerable to blocking of their large-scale foraging movements, loss of secure feeding and breeding grounds, and damage to food trees (especially dipterocarps and oaks).

Sambar deer and barking deer, meanwhile, may have had access to more food in regenerating forest, but were vulnerable to extreme hunting pressure along logging roads and around salt licks. Other species lost feeding and roosting or nesting sites, and many were subject to increased hunting pressure. Arboreal species were particularly vulnerable to increased use of firearms, and nocturnal ones were affected by hunting with the use of firearms and lights. The issue was therefore complex, but the perception that logging was *primarily* to blame for a significant increase in hardship in the rural areas became an important political issue in Sarawak during the late 1980s.

2.1.4 Events in the Baram

The middle and upper Baram river system covers about 22 500 km^2 of northern interior Sarawak and embraces the main Baram river valley and its tributaries and surrounding highlands (Caldecott, 1988a). These include the Tamabu mountain range, with Sarawak's highest peak (Gunung Murud, 2438 m), and the Kelabit Highlands (Fig. 2.2). The area included territory used by several hundred Penans, but most of the population comprised Kayan, Kelabit, Kenyah, and Lun Bawang people of the Orang Ulu group. Perceived conflicts of interest between shifting cultivators and loggers flourished in the Baram from the mid-1980s onwards. This was partly because the longhouses of the upper Baram tended to be surrounded by intact rain forest, while those in the middle Baram had recently had their first experience of widespread logging. Although the area was large and sparsely populated, motorized river transport and government-subsidized air travel had greatly improved communications. This meant that the inhabitants of the upper Baram were able to compare their own life style with that of their downstream neighbours who had already experienced logging.

Because of this contact, longhouse-dwelling people in the upper Baram began to voice their concern about a number of issues. The following comments, for example, were made by members of the Baram public during meetings with the ITTO mission (1990: 165–9):

- 'The first speaker said that most of their land has been destroyed by the activities of logging which leaves timber criss-crossing farmland and fish dying because of water pollution. The animals run away because most of the fruit trees have been knocked down and rattans that are used for

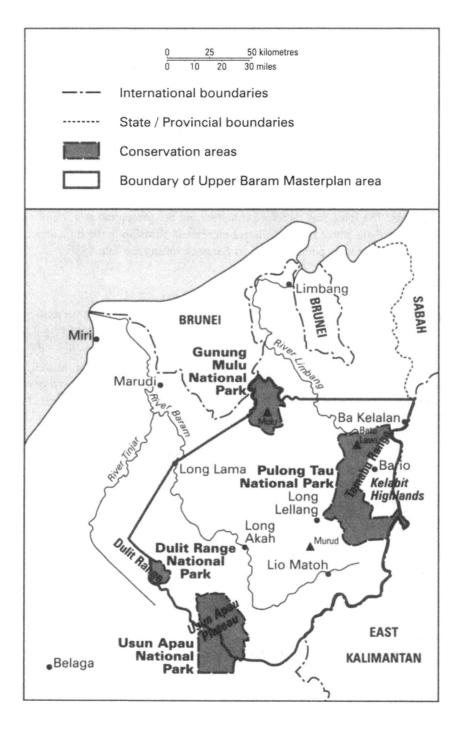

—·—·—	International boundaries	
-------	State / Provincial boundaries	
▨	Conservation areas	
☐	Boundary of Upper Baram Masterplan area	

0   25   50 kilometres
0   10   20   30 miles

Limbang

BRUNEI

BRUNEI

SABAH

Miri

Gunung Mulu National Park

River Limbang

Marudi

River Baram

Ba Kelalan

Batu Lawi

River Tinjar

Long Lama

Pulong Tau National Park

Tamabu Range

Bario

Kelabit Highlands

Long Lellang

Dulit Range

Dulit Range National Park

Long Akah

Murud

Lio Matoh

Usun Apau Plateau

Usun Apau National Park

EAST

KALIMANTAN

Belaga

buildings have also been destroyed, and there is difficulty in looking for food.'

- 'Penghulu Langub said that logging has not yet started in his village but he has seen the effects of logging elsewhere and is anxious that this should not happen when logging comes to his vicinity.'

There was a widespread belief that logging concessions were being issued for the lands around longhouses without their inhabitants being consulted. Moreover, the holders of these concessions were seen as becoming wealthy, while Orang Ulu groups or individuals were not participating fully in these benefits. Environmental damage was also anticipated as a result of logging, which was expected to affect income, lifestyle and diet, and which would damage ritually important sites. It was also thought that monetary compensation for this damage would be inadequate or, if provided in the form of labour fees, inadequately distributed. Lastly, there was a fear of social and religious disruption caused by an influx of outsiders and exposure of young people to outside influences.

A different kind of concern was expressed by the Baram Penans, who in places remained wholly forest-dwelling and without fixed places of residence. Many perceived logging as a direct assault on their way of life and on their idea of environment and their place in it. As Brosius (1993: 142) observed: 'By transforming the landscape, logging destroys those things that are iconic of Penan society. When bulldozers clear roads through the forest, they scour away the surface of the land and obliterate recognizable features ... The cultural density of that landscape – all the sites with biographical, social, and historical significance – is hidden, producing a sort of collective amnesia.'

The circumstances of the Penans were publicized globally during the late 1980s (e.g. Winchester, 1987; Shackleton, 1988). The case was often presented as an example of conflict between the rights of indigenous peoples and the demands of central government and big business. The issue was joined in the minds of many non-Malaysians with similar past and recent conflicts elsewhere, such as the treatment of aboriginal peoples by dominant cultures in Australia, Brazil, Canada, India, Indonesia, the United States, and Venezuela (e.g. Colchester, 1992; Cultural Survival, 1993).

Fig. 2.2 The upper Baram region of Sarawak

For their own reasons, and encouraged by foreigners living among them (of whom the most celebrated was Bruno Manser), the Penans became increasingly militant in their demands for a homeland free from logging. In many instances, the Penans allied themselves with Orang Ulu groups to resist logging by blocking logging roads (Pearce, 1991; Primack, 1991). This direct action was made illegal by the Forest (Amendment) Ordinance of 1987, and the Government also speeded up efforts to help the Penans settle in farming communities. The official position, widely supported by urban Sarawakians, was that forest-dwelling Penans were not *wildlife* and had a right to participate in the mainstream economy which superseded any effect which logging might have on their former life style. In this view, arguments in favour of a continued nomadic life style for the Penans were assumed to come from foreigners hostile both to economic development and to Malaysia.

By the mid-1980s, the scene was set in the upper Baram for a strategic dispute between two main groups with very different attitudes and aims. In summary, many of the Orang Ulu wanted some areas to be protected from logging, and also greater participation in the financial benefits of logging. They were joined for most purposes by some Penans, who wanted a secure area in which to live in a traditional way. The State Government and the timber industry, meanwhile, had an interest in gazetting and using large areas as production forest. Both sides had vociferous allies within the Baram, within Sarawak, elsewhere in Malaysia and in the outside world, and their debate became linked to domestic politics at the State and Federal levels. Global publicity about logging in Sarawak was seen in Malaysia as a threat to the country's tourism industry, while calls for a boycott of Malaysian timber by the foreign press (Clover, 1988) and a hostile resolution of the European Parliament (EC, 1988) were seen as threatening timber exports. As a result, logging became a very heated issue in Sarawak during the middle and later 1980s.

2.2 Designing responses

2.2.1 *Conservation opportunities*

The conservation position in the upper Baram in the mid-1980s could be summarized in the following way. Because large areas had not yet been assigned to permanent land-use categories, there was a general competition for space among groups with special interests in the four divergent priorities of protection, subsistence hunting and gathering, logging, and farming. Meanwhile, because of a lack of research, the role of non-timber forest products (especially wild meat) in the economy of the Baram (and of Sarawak as a whole) could not be quantified, so they were being largely ignored in decisions affecting land use. A lack of research also made it hard to recommend ways to improve the law to protect wildlife populations from excessive hunting or habitat destruction, while maintaining resources needed by rural people.

There was thus the opportunity to document the use of wildlife resources, in order to influence legislation and investment in wildlife protection and forest management. There were also opportunities to create new nature reserves to safeguard the region's environment and biodiversity. On the other hand, there was a sense of urgency since the process of gazetting the PFE in the Baram, and of allocating logging concessions there, seemed likely to be completed by about 1990. This urgency was based on the belief that logging would seriously damage what was then an almost pristine environment, and that it would cause an irreversible loss of biodiversity.

The Baram project was conceived in 1987, following completion of the hunting study and building on its findings and on familiarity with the project area. It developed initially by extending the valuation of non-timber forest products to include items other than wild meat, and sought to justify diversifying the region's economy away from logging. It also sought to encourage protection of key watersheds and habitat areas around the Baram, and to plan for better environmental management in the region as a whole.

2.2.2 The role of non-timber forest products

(a) Gaharu

This is a dark, dense, and deeply fragrant woody material which has been traded internationally for more than 1000 years, being used in perfumes, incense, and medicines in China, India, and Southwest Asia (Burkill, 1966). The best grades were being sold within Sarawak in the late 1980s for US$200–300 per kg. It is produced by certain trees when they are attacked internally by a fungus. The source trees are all members of one family, the Thymelaeaceae, and comprise:

- *Aetoxylon sympetalum*, which is endemic to heath forests in western Sarawak and West Kalimantan, and produces the best or 'true' gaharu;
- *Aquilaria malaccensis*, which is a very rare tree of dipterocarp forests, and produces gaharu of variable quality which is the main source in the Baram; and

- *Gonystylus bancanus*, which is a widespread tree of swamp forests, and produces low-grade gaharu (or *gaharu buaya*).

Harvesting gaharu involves destruction of trees which look as though they might be infected, and this had caused the near-extinction of *Aetoxylon*. It had also caused declining availability of *Aquilaria*, but, despite this, gaharu remained an important forest cash crop throughout the middle and upper Baram in the late 1980s. It was proposed (Caldecott, 1987a) to protect surviving *Aetoxylon* trees in western Sarawak, and to work with Penan communities in the Baram to identify and monitor wild *Aquilaria* trees, while also establishing seedlings and tissue culture in nurseries, and conducting research on the fungal infection itself. The aim was to devise a sustainable harvesting or artificial culture system so as to develop gaharu as a permanent export commodity in Sarawak.

(b) Engkabang

Illipe nuts or engkabang are the seeds of certain dipterocarp trees, especially those of the genus *Shorea*. They contain high-grade oils and are exported to Europe, where they are used especially in the manufacture of Easter eggs and similar moulded chocolate products. The seeds are collected during irregular fruiting seasons by longhouse people, and sold to agents for export (see Box 2.1). This is an important source of cash, and like other fruit trees, engkabang trees are often owned by individuals and families under customary law. The main significance of engkabang is as a supplementary source of income in the context of a forest-dependent farming economy. It was proposed (Caldecott, 1987b) to strengthen the legal and operational protection of *engkabang* trees during logging, while also exploring the possibility of developing a plantation crop by selecting strains which set seed more regularly than once every 3–5 years. This was based on reports of a strain of *Shorea stenoptera* from Java, which fruited annually from a young age (Anderson, 1972, 1975).

(c) Rattans

Rattans or climbing palms belong to the sub-family Calamoideae of the palm family (Dransfield, 1979, 1984). There are about 600 species, most of which occur in the rain forests of South-east Asia and the Malay Archipelago, but which also occur as far away as Fiji and West Africa. By the mid-1980s, the rattans were commercially the second-most important forest product in South-east Asia, after timber, and were used in many products. These especially included furniture, but mats, baskets, fish-traps, and other handicraft items were also important, being needed for subsistence use by farmers. More than 150 000 t of rattan were being traded globally each year, and at least 500 000 people were employed in many countries collecting and processing the canes (Caldecott, 1988b).

A large proportion of raw rattan had come from Indonesia, but this export trade was increasingly restricted during the mid-1980s, causing a widespread shortage of raw material for the manufacture of rattan-based articles. The world price of cane increased sharply, and there started to be interest in rattan as a plantation crop. This was explored on behalf of the Forest Department as an option for large-scale investment (Caldecott, 1987c), and also for longhouse communities on the Baram river (Bennett & Caldecott, 1988). The aim in both cases was to encourage planting in forests which were regenerating after logging, partly to discourage premature and harmful re-entry logging by helping forests to yield a commercial return without the need to cut more timber. Added benefits were thought to include the way in which local communities could use canes to produce higher-value finished products locally, thus increasing the local capture of economic value from the crop.

(d) Other potential products

Non-timber products from rain forests include chemicals which it was thought might yield pharmaceutical products, and this possibility was explored with attention to applications in medicine and dentistry (Caldecott, 1987d, 1987e). The aim was to draw attention to a range of new options for using forest resources. These were brought together into a proposal to establish a Natural Products Institute (NPI) in Sarawak (Caldecott, 1988c). The role of the NPI was to be to develop commercial products from among the full range of native species in Sarawak, while also using the ethnobiological knowledge of the people of Sarawak's interior to focus attention on potential pharmaceuticals, in return for their sharing in profits.

A model was also developed with which to modify management of forests in Sarawak (Caldecott, 1988d). This was based on work by Tang (1987), who had previously concluded that it would be more efficient to manage forests for timber on a stand-specific basis, taking into account the varied composition and structure of real forests, rather than applying a single harvesting régime to all forests. The modified proposal was to manage forest concessions through detailed pre-felling inventories, allowing each stand to be used according to its most promising attribute, of which timber was to be treated as only one of many options. Others were to include wild meat, tourism, water supplies, rattans, fruits, and chemical products. Where logging was chosen, the pre-felling inventory would allow trees to be individually selected and marked for directional felling, thus reducing damage to the residual forest stand.

2.2.3 The role of rhinos

A proposal to create Pulong Tau National Park in the Tamabu Range was prepared by Mike Kavanagh of WWF Malaysia for the Forest Department in 1984 (NPWO, 1984). The aim was to safeguard a large sample of middle to high altitude forests and steep watersheds between the Kelabit Highlands and the middle and upper Baram River. By 1986, however, this proposal had been shelved because of the increasingly public debate about logging, and because of uncertainty concerning the boundaries of logging concessions which might abut or overlap the boundaries of the proposed Park.

The 1984 proposal mentioned that a population of the Sumatran rhinoceros (*Dicerorhinus sumatrensis*) probably still existed in the area. This species is among the most endangered large mammals in the world, and had been thought to be extinct in Sarawak since the 1950s (Harrisson, 1956). Based on Penan reports of tracks in the northern zone of the proposed Park, there was little doubt that some rhinos survived, although the State Government was unconvinced that these were not vagrant individuals entering Sarawak from Kalimantan. If it was confirmed that a resident population existed within the Park, it seemed possible that this would prompt rapid gazettement of the Park.

Resources were therefore obtained from the Earthlife Foundation, with which to survey the headwaters of the Limbang river, around the base of the spectacular monolithic mountain Batu Lawi. This is an area which is almost never visited by Kelabit hunters because of a high mountain wall between Batu Lawi and the Kelabit Highlands, and it was very remote from the low-lands before logging roads extended up the Limbang valley. Because of this isolation, in 1986 the wildlife of the area still showed little fear of people, and the presence of several rhinos was confirmed (Caldecott, 1987f). These findings implied a resident population living in one of the most inaccessible parts of the proposed National Park.

The Forest Department therefore requested a revised proposal for the National Park in July 1987 (NPWO, 1987a). The Park was to be about 1600 km^2 in area, with some 80% in the Baram District and the rest in the Limbang headwaters. It was to include Gunung Murud, Gunung Batu Lawi, the area known to be occupied by rhinos, oak forests used as feeding and breeding grounds by migratory bearded pigs, and the Tamabu Range which comprises much of the catchment of the Baram river system. The proposal also emphasized that there was very strong local support for the Park, and that the name *Pulong Tau* was suggested locally and means *Our Forest* in the Kelabit and Lun Bawang languages.

The next task was to prepare a management plan for Sarawak's rhinos based on the assumption that a large (320 km²) core area around Batu Lawi would be protected within the Park, and that access to logging areas downstream in Limbang could be carefully controlled by the Forest Department to prevent poaching. In these areas, outside the Park, entry to logging coupes was to be scheduled and managed to ensure that rhinos if present were displaced towards the core area, rather than away from it (Caldecott, 1987g). Meanwhile, Kelabit, Lun Bawang, and Penan communities enthusiastically accepted responsibility from the Forest Department for denying access to poachers on their lands (Caldecott & Labang, 1988). This local support came largely from the belief that the rhinos would justify the Park, which would protect the environment of the Kelabit Highlands and the upper Limbang valley, and ensure continued harvests of irrigated rice and wild bearded pigs.

The Wild Life Protection (Amendment) Ordinance of 1988 increased penalties for rhinoceros hunting to two years' gaol and a fine of over US$10 000, and this fine was more than doubled again in the new law of 1991. The Forest Department in 1987–8 seemed confident that Pulong Tau National Park would be gazetted as proposed, even though it was rumoured that the rhino core area had already been included within a logging concession. It was thought that the presence of rhinos would make it feasible to relocate the concession, since all parties accepted that there was no possibility of preserving the rhinos if timber workers were allowed access to the core area. Although this outcome was not in the end realized (see below), at the time it seemed that Pulong Tau National Park and the rhinos were secure.

The Forest Department then requested proposals to protect other sections of the Baram river system within National Parks, and this was done for Usun Apau and the Dulit Range, on the southern and western boundaries of the Baram project area (NPWO, 1987b, 1987c; Box 2.2). Together with the proposed Pulong Tau National Park and the existing Mulu National Park (Jermy & Kavanagh, 1982, 1984), the four areas were intended to secure key environments in an arc around the middle and upper Baram (Fig. 2.2).

Box 2.2 Key features of the proposed Pulong Tau Usun Apau and Dulit Range National Parks (from NPWO, 1987a, 1987b, 1987c)

Pulong Tau National Park was proposed to be 1590 km² in area, and to include: the peak of Gunung Murud, Sarawak's highest mountain; Gunung Batu Lawi, a spectacular monolithic landform; forests occupied by the Sumatran rhinoceros, one of the world's rarest animals; oak forests used as feeding and breeding grounds by migratory wild pigs, which are hunted for food throughout northern-interior Borneo; and the Tamabu Range, including much of the catchment of the Baram river system.

Usun Apau National Park was proposed to be 1130 km² in area, and to include: the Julan Falls, Sarawak's most spectacular waterfall complex; the Eastern Plateau of Usun Apau, a high volcanic tableland bounded by tall cliffs; a variety of unique ecological transitions between the plateau mossy forests and the rain forests of surrounding river valleys; an example of unusual swampy forest; sites of major archaeological significance and culturally important to the Kenyah people; and much of the catchment of the Baram, Tinjar, and Murum-Balui river systems

Dulit Range National Park was proposed to be 140 km² in area, and to include: a series of ten peaks between 1200 and 1500 m elevation, which together mark the southern end of the Dulit Range; some of the most precipitous terrain in Sarawak, including *ca.* 40 km of pure white sandstone cliffs; a series of viewing points from which spectacular scenery over 150 km of northern Borneo can be seen; a significant part of the rich Dulit Range ecosystem; and the habitats of several species which are rare or extinct elsewhere, including Hose's civet (*Hemigalus hosei*).

2.2.4 The Baram masterplan

The Forest Department encouraged the study of options to diversify the State's forestry investments, and in 1988 it became possible to develop a proposal for a comprehensive planning exercise for the middle and upper Baram. This would take account of opportunities to develop tourism, wildlife and fisheries management, rattan plantations and new forest crops, and allow for low-impact logging and plantations of indigenous tree species in suitable areas (Caldecott, 1988e). The aims of this proposal were similar to those of the proposed Baram Hinterland Feasibility Study which had been considered and shelved by the State Government in October 1984. This earlier proposal was presumed to have been unsuccessful because of the increasing sensitivity of land planning issues in the Baram. It had sought to review the total environment of the Baram, to prepare maps of land use and land suitability, and to develop ideas for local participation in development, while protecting important environmental assets and diversifying the local economy (Box 2.3).

Box 2.3 The aims of the Baram Hinterland Feasibility Study, as considered by the Sarawak Government in October 1984

(a) To review the physical, biological, cultural, and economic environment of the Baram area, mapping land-use and land suitability, and assessing infrastructure, settlements, navigable rivers, and the local road system.

(b) To conduct socioeconomic surveys and develop ideas for local participation in development while maintaining the cultural heritage of the area.

(c) To identify forest land to be protected for environmental and watershed use or as parks and sanctuaries.

(d) To identify areas suitable for agricultural projects and select appropriate crops and projects, including use of temperate agriculture and livestock in the highlands, and riverine aquaculture (especially of the fish *Tor tambroides*).

(e) To assess opportunities to develop local timber-processing and small hydro-electric facilities.

(f) To assess opportunities to develop tourism in connection with Mulu National Park and the Baram River Club [later part of the proposed Usun Apau National Park], and to recommend ways to maintain local traditions while promoting tourism.

(g) To conserve caves used by edible-nest swiftlets (*Collocalia* spp.), and their populations, while promoting sustainable production of nests.

(h) To identify the main constraints on development, especially relating to communications and transport, and make recommendations to relieve them.

(i) To analyze and evaluate the market potential of all products and services derived from the region.

(j) To examine the need for land consolidation through land settlement operations and the issue of land titles.

(k) To review critically all existing development policies and projects for the area.

The 1988 proposal had themes which were similar, but up-dated in the light of later work on non-timber forest products and the proposed system of national parks in the Baram. It also envisioned the possibility of using investment incentives within the Baram to encourage development of new forms of sustainable economic activity based on the area's biodiversity (Caldecott & Mundy, 1987). One option to be considered was the use of reciprocal taxation agreements between Malaysia and other countries, whereby losses on investments in Malaysia could be used to reduce tax burdens in the investor's home country. This mechanism had previously been used to assist British investment in forestry in the United States and Canada. Another option was to negotiate the use of sovereign debt owed by the Government of Malaysia to other governments, converting it into low-interest financing for certain categories of investors. Other investment incentives such as tax holidays and direct subsidies were also to be explored, as ways to encourage long-term investment in the Baram by the private sector.

This proposal received a mixed reaction from Government, with high-level support from some quarters and reservations from others. It was opposed, and eventually vetoed, by the Ministry of Land Development because of anxiety over allowing foreigners to work in such a sensitive field as environmental management in the Baram. By that time, June 1988, it had emerged that logging concessions had in fact been issued for key portions of both the Pulong Tau and Usun Apau proposed National Parks, including the rhino core area in the former, and most of the important wildlife habitats in both, and that these concessions were not going to be revoked or relocated. These factors led to proposals for further work on the Baram masterplan being shelved.

2.3 Discussion

In Sarawak, and particularly in the Baram, there was an opportunity in 1984–8 to help increase the area of rain forest to be set aside for protecting biodiversity and environmental services. Some progress was made by strengthening the economic case for protecting forests, mainly by documenting the value and potential of non-timber forest products such as wild meat. There was also the opportunity to help the Forest Department formulate national park proposals and legislation. All of this work ultimately contributed to a more ambitious project aiming to help achieve in the Baram a balanced compromise between the needs of the forestry sector, wildlife, and local people while keeping as much species-richness as possible within protected and well-managed forests.

The Sarawak Forest Department shared these aims to a large extent, since it maintained a strong interest in protecting biodiversity and in managing forests to high professional standards, with minimum incidental environmental damage. The Forest Department, however, and all conservationists working with it, were in a difficult position during this period. This was because the role of the Forest Department was to supervise a forestry system which functioned by awarding and exploiting industrial logging concessions for profit, in order to support the State's growing economy. The demands of this system on natural resources were severe, and conflicts naturally arose between these and the needs of wildlife and local communities within the areas affected. Everyone concerned had therefore to maintain a delicate balance, which became increasingly difficult to sustain as the issue of logging in the Baram became politicized to the point where powerful interest groups began to feel themselves threatened.

At times, these tensions seemed likely to make any progress impossible, but in the end the Sarawak Government did succeed in reducing overall logging rates, while reforming the wildlife law and continuing to process National Park proposals which should secure parts of the Tamabu Range and the Usun Apau plateau. The interior of the Baram will never again be as pristine as it was in 1984, and the opportunity to manage the area in an 'ideal' manner was lost for ever with the shelving of the Baram Hinterland Feasibility Study proposal in that year. Nevertheless, many of the species native to the Baram region will survive within those protected areas which have been or will be created, and in small relict forest stands in inaccessible locations. The area has the great advantage that there are so few people waiting to colonize and plant crops within the logged forests. This implies that, unless forest fires intervene, the landscape will be dominated by natural forests into the indefinite future.

3

Cross River, Nigeria

3.1 Conservation issues

3.1.1 Land and people

At about 911 000 km^2, Nigeria is about the same size as Venezuela and is larger than New Guinea, Borneo, or Texas. It lies between the southern margin of the Sahara Desert and the Atlantic Ocean, bordered by Niger and Chad to the north, Benin to the west and Cameroon to the east (Fig. 3.1). With nearly 90 million inhabitants in 1990, it is the most populous country in Africa and has about 14% of Africa's whole population. Hundreds of tribal groups live there, with many religions of which Islam and Christianity are the most common. Nigeria may be described simplistically as being divided into three parts, the north, south-west and south-east. These are dominated respectively by the mainly Muslim Hausa-Fulani, and the mainly non-Muslim Yoruba and Igbo peoples. These groups can be viewed as competing for national power, with the minority groups of the Middle Belt having a balancing role. This crude analysis does suggest how rulers have been able to transcend and exploit tribal and religious divisions, and to balance rival factions within first a Colonial and then a Federal political structure influenced by long periods of military rule.

In the later Colonial era, Nigeria was ruled by the British through the three logical Regions of North, West and East. Following independence in 1960, however, there was a move to create a Federal system based on states. A number of states were created in 1967, but this upset the traditional balance of power among major regional interests and led to the Civil War of 1967–70 (Forrest, 1993). This war involved the Eastern Region in an unsuccessful attempt to establish itself as independent Biafra. It was a major national trauma that continues to affect official and public attitudes towards movements for increased regional autonomy. The latter are influenced by complex

30

Design based on a pre-historic South-east Nigerian stone goddess.

relationships among the dominant group and smaller ones in each major part of the country, for example the tensions between the Igbo and the Ogoni, Andoni, Ikwere, Ijo, and others in the Niger Delta (N. J. Ashton-Jones, personal communication, March 1994).

The victors of the Civil War continued to build a Federal Nigeria based on increasing numbers of states: 12 in 1967, 19 in 1976, 21 in 1987, and 30 in 1991, plus the Federal Capital Territory of Abuja. The Federal structure allowed the states a high degree of political autonomy, but few opportunities to raise their own revenues. The states have therefore relied on money from the Federal Government, which has largely come from the sale of oil. This meant that the states were encouraged to compete for Federal funds, and this had several important effects. At the local level, there was an incentive to create new states, which could give new groups access to Federal funds. At the state level, there was an incentive to maximize current expenditure in

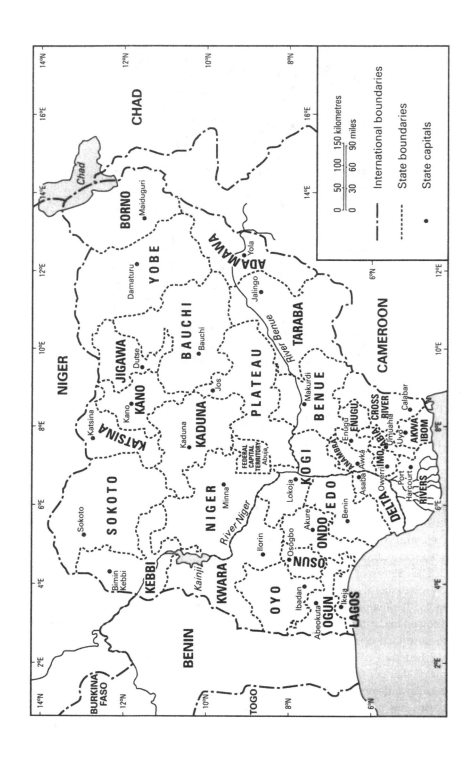

order to reward local political loyalties. At the Federal level, there was an incentive to maximize current oil revenues, which could be used to reward the continued political loyalty of the states themselves.

These arrangements locked the Federal and state governments into a difficult relationship from which the former has sought a way out since the late 1970s. This involved strengthening a tier of government below that of the states, the Local Government Areas (LGAs). Initial reforms moved these away from being a tier of traditional or appointed authority, and towards being governed by councils with elected participation. They were to be given an increasing degree of autonomy by the states, and were initially granted 10% of Federal expenditure. Devolution of power to the LGAs ground to a halt during the early 1980s, and the LGAs remained effectively under state control, largely because their Federal allocations were paid at the discretion of the state governments.

The Federal Military Government from 1986 to 1993 sought to restore momentum to the process of LGA empowerment, partly by increasing their share of Federal expenditure from 10% to 15%, and then in 1992 to 20%. The LGAs were given responsibility for primary education and health care, and Federal grants were paid directly, along with a supplement of 10% of the states' internally generated revenues. A number of other reforms all tended to strengthen the powers of the LGAs (Forrest, 1993). The aim of these changes was apparently to transfer power away from state institutions, thereby reducing the importance of the state level of government and changing the nature of state–Federal relations. This process was resisted by groups who perceived it as a threat to their own positions. Tensions arose which led firstly to the creation of yet more states in 1991, and then in 1993 to the abrupt restoration of military rule following national elections that were afterwards annulled. These events reconfirmed military dominance over the Federal government.

An effect of the devolutionary policies of 1986–93 was to create smaller and more tribally homogenous states and LGAs. This was significant for conservation, since it brought new, and in some cases more sympathetic, groups into positions of influence (see below). Another factor was the publication of the National Parks Decree No. 36 of 1991 (FGN, 1991), which formalized the role of the national parks as the main official means for con-

Fig. 3.1 Nigeria

servation in Nigeria, and established procedures for them to be administered and financed. In effect, national parks became recognized as vehicles for Federal resources to be allocated to those states (and LGAs) which possessed them. They thereby became a new kind of mechanism for balancing the interests of the Federal, state, and local tiers of government, while also satisfying the immediate needs of an increasingly influential group of people concerned about conservation.

3.1.2 Environments and biodiversity

Nigeria's climate ranges from very dry in the north, where some places have less than 500 mm of rain a year, to very moist in the south, where annual rainfall exceeds 2000 mm and in places 4000 mm (WCMC, 1988; ODNRI, 1989). There is a single wet season of 90–150 days in the north (in July and August), a bimodal wet season of 200–300 days in the middle part of the country and the inland south (with peaks in June and September), and year-round wet conditions in the extreme south (with a drier season at the turn of the year). Variations on these climates are found in highland locations in the centre of the country and on the Cameroon frontier.

There is a corresponding range of natural vegetation types, from desert and Sahelian scrub in the north, through northern and southern Guinea savannahs in the Middle Belt, to moist deciduous forests, mangroves, and rain forests in the south (Sayer, Harcourt, & Collins, 1992). Nigeria is divided roughly into three by the Niger and Benue rivers, which join before draining southwards into the enormous wetland of the Niger Delta on the Atlantic coast. This Delta is more than 20 000 km^2 in area, and comprises an inland floodplain and extensive mangrove swamps behind coastal barrier islands (Wilcox & Powell, 1985; CEC, 1987).

Pressure on Nigeria's natural resources is intense in many parts of the country. This is reflected in the status of the forests, with deforestation and desertification proceeding rapidly in the north, and widespread mismanagement and inappropriate conversion of moist forests having occurred in the south (Areola, 1987; Caldecott & Fameso, 1991). In the 1890s, Nigeria was estimated to have about 600 000 km^2 of forest, but this had declined to about 360 000 km^2 in the 1950s, and by the 1980s only about 140 000 km^2 remained, of which less than half had a closed canopy (Sayer, Harcourt, & Collins, 1992). The recent history of Nigeria's environment has been catastrophic, marked by a lack of investment in rural environmental management, and by the deliberate or accidental abuse of all forms of renewable natural resource. As the Sarawak Director of Forests fairly observed, Nigeria's 'problems on forest conservation are far greater than ours' (Datuk Leo Chai, personal communication, January 1989).

If recent trends continue, the population of Africa will increase from about 654 million in 1992 to 3090 million in 2025 (UNFPA, 1991; Dompka, 1994). Under these conditions, Nigeria's environment in the 1990s may represent what much of Africa will be like within a few decades (WCMC, 1988). Solutions to environmental problems developed in Nigeria may therefore be applicable in many other locations. Time to develop such models is likely to be short, however, since Nigeria has already shown a high rate of species extinction and widespread degradation of protected areas. Its own population, moreover, is expected to increase by several hundred millions in the same time period.

Extensive recent deforestation in Nigeria has had a profound impact on all aspects of the country's environment. This environment is extremely varied, and dominant threats vary from place to place. A World Bank study (Singh *et al.*, 1990) condensed them into the three key national problems of soil degradation, water contamination and deforestation. It also noted other severe and widespread problems, including gully erosion, damage to fisheries, coastal erosion, loss of wildlife and biodiversity, and air pollution. A survey of the country by Caldecott & Fameso (1991), highlighted the following key problems:

- fire degrading moist forests to savannah woodlands, and to further damaged states;
- fuelwood collection for subsistence or commercial use;
- browsing and grazing pressures by the livestock of both sedentary and nomadic peoples;
- lopping or felling of trees for fodder;
- illegal hunting inside and outside nature reserves;
- conversion of forest land for both large-scale and small-scale farming;
- failing floods in floodplain areas;
- ephemeralization of rivers;
- poorly planned, poorly managed, and poorly supervised large-scale and small-scale timber exploitation;
- soil erosion by wind or water;
- pollution from various sources; and
- urban environmental problems in great variety.

Southern Nigeria was once almost entirely covered in moist forests, with very large areas of mangrove in the Niger Delta and other estuaries, and rain forests grading northwards into increasingly seasonal moist forests and southern Guinea savannah. These natural moist forests are now largely fragmented and discontinuous, partly because of widespread conversion to forest plantations, partly because of population growth and farming, and partly because of fire damage causing derived Guinea savannah to spread southwards into the moist forest zone. Nevertheless, some large areas of moist forest remained in 1994, in particular in the Niger Delta, where up to 4000 km² of mangrove remained intact although intensively used by people, and other large areas of fresh-water swamp forest survived in a moderately disturbed state further inland (N. J. Ashton-Jones, personal communication, March 1994; J. F. Oates, personal communication, April 1994). There were also significant areas of natural moist forest to the north of the Niger Delta, and in the extreme south-east of the country.

Nigeria includes or overlaps several biogeographical units, including the rich Guinea-Congolian forests which cross the eastern frontier into Cameroon (Sayer, Harcourt, & Collins, 1992). These represent both a centre of plant diversity (WWF & IUCN, 1995) and an endemic bird area (ICBP, 1992, 1995). National species richness is therefore high, with over 4600 plant species, and about 250 mammals and 840 birds. Many of these species are concentrated in the extreme south-east of Nigeria, which is believed to have been a moist-forest refuge during Pleistocene Ice Ages, and which has about 3500 plant species and most of Nigeria's endemics. This area is biogeographically highly distinctive, and many species are confined or almost so to the region ('Oatesia') between the Sanaga River in Cameroon and the Cross River in Nigeria. These include many amphibians and three monkeys, a forest baboon (the drill, *Mandrillus leucophaeus*) and two guenons (*Cercopithecus erythrotis* and *C. preussi*). Apart from the butterflies (Boorman & Roche, 1957–1961; Collins & Morris, 1985) the invertebrate fauna is little known, but is thought to be extremely rich.

Intensive and widespread farming is particularly marked in south-eastern Nigeria, where population densities are up to 1000 people km⁻², and many species are under threat. In Imo, Anambra, Enugu, and Abia States, for example, over 4% of the land area was affected by serious erosion by 1990 (Singh *et al.*, 1990), and in 1991 these states together possessed about 1200 major gully erosion sites (Caldecott & Fameso, 1991). In nearby Ogoniland, further west, it was estimated in 1994 that severe sheet erosion affected about 40% of the land area (N. J. Ashton-Jones, personal communication, March 1994), and species-rich lowland rain forest had been mostly replaced by an artificial monoculture of oil palms (*Elaeis guineensis*).

3.1.3 Cross River State and its forests

Cross River State (CRS) occupies the land between Benue, Enugu, Abia, and Akwa Ibom States of Nigeria and the Cameroon frontier, on either side of the Cross River (Fig. 3.2). Its natural environments extend from the mangroves of the Cross estuary, through rain forests in the south and centre of the State, and into the southern Guinea savannah zone. In 1990, in contrast to much of the rest of south-eastern Nigeria, population densities were low, and most agriculture was concentrated in the Cross River valley and the raised coastal terrace. Rugged or swampy terrain elsewhere had proved unattractive to farmers, and this helped to limit the growth of communications and transport links with the rest of the country. These factors tended to prevent excessive pressure on the natural forests, and through the late 1980s the State continued to possess some of the largest and most intact areas of tropical moist forest still left in Nigeria.

This feature attracted the attention of the World Conservation Union (IUCN), which in three Africa-wide studies identified the forests of CRS as worthy of special conservation measures (MacKinnon & MacKinnon, 1986a; IUCN, 1987a, 1987b). All three studies emphasized the extreme biological richness of the surviving forest resources, their uniquely intact status, and the increasing threat to their integrity represented by uncontrolled farming, logging, and hunting. This international attention was based on a long history of interest in Cross River's forests and wildlife, concentrated on the Oban forests to the south of the Cross River and the Boshi-Okwangwo forests to the north. Proposals for conservation action date back to the 1950s and 1960s (Box 3.1), but work was delayed by the Civil War and its aftermath. Proposals were renewed in the 1980s by Cross Riverian biologists, and then by field staff of BirdLife International (ICBP), the Nigerian Conservation Foundation (NCF) and the World Wide Fund for Nature (WWF). Surveys by WWF and NCF in 1988 concluded that a conservation project in CRS would be both desirable and feasible.

Box 3.1: History of attention to potential nature reserves in Cross River State (CRS) of Nigeria (adapted from Caldecott, Oates, & Ruitenbeek, 1990)

1956 E. W. March of the Government of the Eastern Region of Nigeria initiates the Boshi Extension Forest Reserve, mainly as a gorilla sanctuary.

1964 The Eastern Region and Federal Nigerian Governments propose the Obudu Game Reserve for the Boshi, Boshi Extension, and Okwangwo Forest Reserves.

1965 G. A. Petrides recommends national park status for the Obudu Plateau and the Oban Forest Reserve.

1968 J .F. Oates recommends wildlife and habitat protection programmes to be carried out in the Obudu and Oban areas after the Nigerian Civil War.

1979 D. Poore suggests Boshi-Okwangwo as a World Heritage Site.

1981 J. B. Hall argues in favour of conserving the Okwangwo-Obudu ecocline within a national park.

1983 C. O. Ebin of the CRS Government surveys the proposed Obudu Game Reserve, confirms the presence of gorillas in the Mbe Mountains, and advocates the area's inclusion within a Boshi-Okwangwo National Park.

1986 J. S. Ash & R. E. Sharland of BirdLife International (ICBP) advocate conserving the Obudu Plateau, Boshi-Okwangwo, and Oban areas.

1987 March: J. S. Ash of ICBP proposes national park status for the Boshi-Okwangwo and Oban areas.

1987 July: J. H. Mshelbwala of the Nigerian Conservation Foundation (NCF) reports on the status of the Boshi-Okwangwo gorilla population.

1987 December: A. H. Harcourt, K. J. Stewart, & I. M. Inahoro of NCF survey the Boshi-Okwangwo and Afi River areas, and advocate gorilla sanctuaries, integrated rural development, transfrontier co-operation with Cameroon, and gorilla-based tourism.

1988 January: L. J. T. White & J. C. Reid of NCF argue for conserving the Oban Hills and Ikpan Blocks of the Oban Group Forest Reserve.

1988 April: J. S. Gartlan & C. M. Wicks of the World Wide Fund for Nature (WWF-UK) propose the Oban National Park concept to the Nigerian authorities.

1988 May: CRS Government approves the constitution of a national park in the Oban and Boshi-Okwangwo areas.

1988 June: F. S. Sullivan of WWF and I. M. Inahoro of NCF prepare a preliminary description of the Oban Park area.

1988 September: E. L. Gadsby & P. D. Jenkins of NCF begin managing the Gorilla Conservation Project in the Mbe Mountains, Okwangwo area.

1989 February: HRH Prince Philip (WWF International President) visits Oban and Okwangwo areas to inaugurate the National Park feasibility study and to demonstrate global interest in conservation in CRS.

1989 December: Oban Division feasibility study completed by WWF and the Overseas Development Natural Resources Institute (ODNRI), and accepted.

1990 September: Okwangwo Division feasibility study completed by WWF and accepted.

1991 August: Decree No. 36 published defining Cross River National Park, including both Oban and Okwangwo Divisions.

1993 October: European Union-financed management contract awarded to develop Cross River National Park and its support zone.

Proposals for a national park based on these surveys were welcomed in 1988 by the CRS Government. This response was made possible partly because in late 1987 the State had been divided into two. The south-western third had become a new Akwa Ibom State, while the rest remained as Cross River State. Akwa Ibom was one of Africa's most densely populated areas, and accounted for more than two-thirds of the population of the former CRS. Most of these people were Ibibios and this group hardly occurred elsewhere, so dividing CRS greatly increased the direct and indirect influence of non-Ibibio tribal groups in the CRS Government. These especially included the Ejagham, Korup, and Boki people who lived in rural areas around the forests. Since the proposed Park was presented as a project for the benefit of these local people, their increased influence was exerted strongly in support of the project.

The involvement of WWF in CRS in early 1988 had its background in the Korup project in neighbouring Cameroon. This was started by the Earthlife Foundation, with joint funding by the British Government, but from 1987 it came under management by WWF. The forests of Korup National Park are contiguous with those of the Ikpan block of the Oban area in Nigeria, and WWF saw the chance of protecting some of the Nigerian forests as well. Proposals by WWF for work in CRS affected two areas, separated by some 40 km and the Cross River itself. The southern (Oban) Division of the Park contained about 2800 km² of rain forest in the Ikpan and Oban blocks, while the northern (Okwangwo) Division comprised about 920 km² of more seasonal forest. The two areas thus complemented one another in possessing different natural ecosystems. Both Divisions were included within the National Park when it was constituted by Presidential Decree in 1991.

3.1.4 The Oban Division

The Oban project area lay north-east of the CRS capital, Calabar, inside a loop of the Cross River, and within the Local Government Areas of Akamkpa and Ikom (Fig. 3.3). The Ikpan block extended for about 40 km along the Cameroon border, where it was partly contiguous with the Korup National Park. The only major highway within the project area was the MCC road, which bisected it from Calabar to Ekang on the Cameroon border. This road passed through Oban, the largest settlement in the immediate vicinity of the National Park, and the junction of another, incomplete road, connecting Oban to Akamkpa. The MCC road had attracted ribbon settlement in a corridor between two hilly and mountainous regions that comprised the core areas of the Park, the Oban Hills to the west and the Ikpan Block to the east.

The two forest blocks approached one another most closely near the Cameroon frontier, and on several stretches of the road the forest canopy

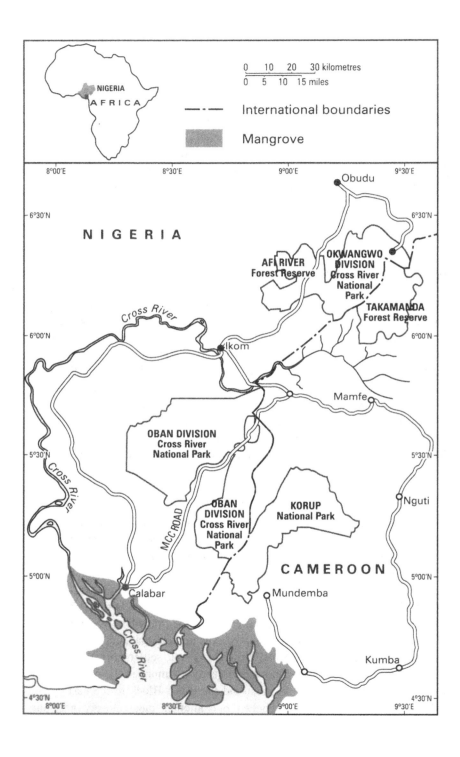

arched overhead from both sides. This meant that there remained at least the possibility of dispersion across the road by flightless, arboreal species, and the proposed Park boundaries reflected this by crossing the road between village lands. It was at one point proposed to join the forest canopies on either side of the road by means of wire cable 'chameleon bridges', which would also be used as an educational resource for travellers on the MCC road and nearby residents.

Terrain in the project area rises from the encircling river valleys to over 1000 m within the Park on both sides of the MCC road. These mountainous areas drain into the Cross River to the north and to the Calabar, Kwa, and Korup Rivers to the south, with few streams running to east or west. The soils, derived from ancient metamorphic rocks, are sandy and infertile, and stony, shallow, and erodible on steep slopes. Average rainfall decreases from south to north across the region, but is affected locally by altitude, so that the central parts of the Park receive an average of over 3500 mm, which is more than in Calabar, near the coast to the south. The rains generally peak from July to September, but the timing varies among localities, and there is a dry season from November to March or April, coinciding with the *harmattan*, a dust-laden dry wind from the Sahara Desert.

Vegetation in the project area had in many places been modified by people, but intact rain forest remained in areas of more rugged terrain. These forests were mostly within the Park, but also extended onto gentler slopes nearby, where most had been logged or converted into a patchwork of small farms, tree plantations, and regenerating forest. Around villages, permanent and semi-permanent farming occurred, with a strong traditional agroforestry component (Okafor, 1979), while elsewhere oil palm and rubber estates had been developed.

3.1.5 The Okwangwo Division

The Okwangwo project was located south-east of Obudu Town, within Obudu (later Obanliku) and Ikom (later Boki) Local Government Areas. A metalled highway connected Calabar with Ikom, where it bridged the Cross River and continued to Obudu Town. Another surfaced road connected

Fig. 3.2 Cross River National Park

Obudu Town with the Obudu Cattle Ranch on the Obudu Plateau. The project area was thus encircled by the Cross River to the south, the Ikom–Obudu Highway to the west, the Obudu Town–Cattle Ranch road to the north, and the Cameroon frontier (Fig. 3.4). Minor and unsurfaced roads penetrated the project area from the main roads, and approach the Park in several locations. The Okwangwo Division was to comprise the whole of the Okwangwo, Boshi, and Boshi Extension Forest Reserves (about 70% of the area), together with all of the Obudu Plateau (11%) and the communal lands between the Okon River and the Ikom–Obudu Highway known as the Mbe Mountains (10%). It was also to include the western part of the Balegete Enclave and all of the Okwa and Okwangwo Enclaves which exist within the Okwangwo Forest Reserve (9%). It was to have a total area of about 920 km² and an external perimeter of about 200 km.

Terrain in this area rises northwards and eastwards from the surrounding lowlands, from about 150 m elevation to over 1700 m on the Obudu Plateau. The soils are similar to those of the Oban area, but the topography is more complex, with many disjunct and connected ridge systems, isolated peaks and outcrops. Rivers in the area generally flow from north-east to south-west, but the complex landform at higher altitudes causes the direction of river courses often to change abruptly. The three main rivers are the Oyi, Okon, and Afi, all of which flow into the Cross. Average rainfall decreases from north-east to south-west across the project area, since the high land along the Cameroon frontier intercepts rain-bearing southerly winds between May and November. The eastern and north-easternmost parts of the project area receive up to 4500 mm of rainfall annually, and experience dry seasons of only two or three months. The Obudu Plateau itself, however, has a lower average rainfall and a more extended dry season, but its high elevation means that the weather there is often cool and misty. The Afi River valley nearby receives less than 2500 mm of rain annually with a four-month dry season, and Obudu Town is both drier and more seasonal than that.

3.1.6 The Oban and Okwangwo Divisions compared

Pressures on the Oban and Okwangwo Divisions of the Park were similar in some ways, and different in others. Some of their main features in 1988–9 are compared below.

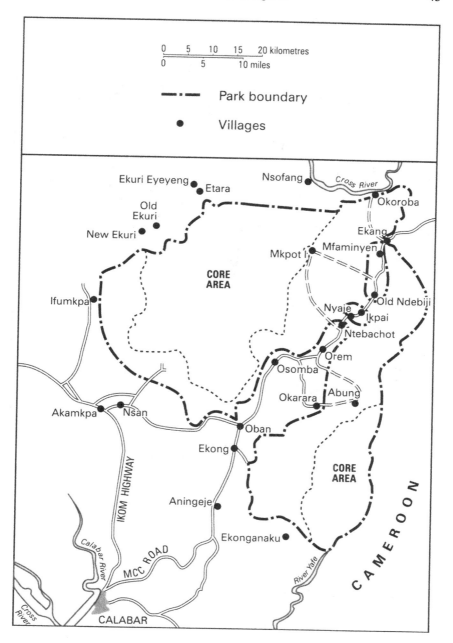

Fig. 3.3 The Oban Division of Cross River National Park

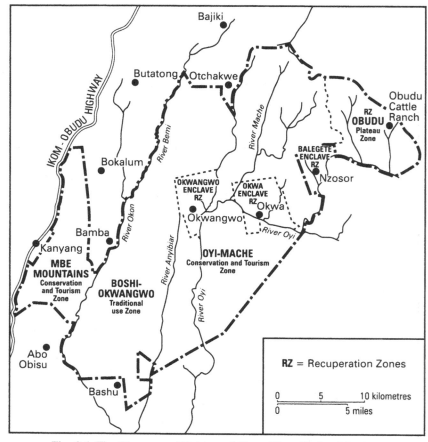

Fig. 3.4 The Okwangwo Division of Cross River National Park

(a) Logging and plantations

In Oban, poorly planned and virtually unsupervised logging had occurred in
all the forest reserves concerned, and other large areas near the Park were
being converted to pulpwood plantations. There was less commercial logging
in Okwangwo, partly because of poor access, and partly because the forest
contained fewer large trees of species with valuable timber.

(b) Farming

The forests in both project areas were under pressure from farmers, with about 40 000 people living close to the Oban Division and about 36 000 around the Okwangwo Division. These people were colonizing increasingly steep land in order to grow traditional crops to feed expanding village populations, and to take advantage of high food prices within Nigeria caused by national macroeconomic policies after the mid-1980s, which included significant devolution of the Naira (Chapter 8).

(c) Fire

Much of the Okwangwo Division had a climate both drier and more seasonal than Oban's, and the Okwangwo forests were therefore in places and at times more vulnerable to fire. Fire was used routinely by local people to clear land for agriculture, and sometimes spread into the forest because of poor supervision. The vegetation in and around the Okwangwo Division showed a complex pattern of exposure, resistance, and adaptation to fire, reflecting its location in a zone of tension between dominant moist savannah and rain forest vegetation types, combined with extensive human influence. Although not widespread, fire was degrading natural ecosystems in places, and there was clear potential for further damage.

(d) Villages within the Park

In the Oban Division, villages along the MCC road had already virtually divided the Park area into two parts, and several were surrounded by Park lands as a result of the Park crossing the road. Two small villages were also located deep within the Oban Division. In the Okwangwo Division, a line of village enclaves crossed the Park area, and the continued expansion of their farmlands threatened to fragment it. These factors meant that the project had to try to find realistic and acceptable ways either to stabilize the use of forest lands by local people, or to resettle several communities.

(e) Hunting

In both project areas, hunting and trapping sustained an important trade in wild meat which provided as much as a third of household incomes as well as satisfying subsistence needs. The extent of intact forest was much greater in the Oban Division than in Okwangwo, however, implying generally larger wildlife populations which would be less vulnerable to extinction. Another factor was that several endangered mammals were present in the Okwangwo Division but not in Oban, including West Africa's only gorillas (*Gorilla gorilla*), and Preuss' guenon (*Cercopithecus preussi*; Oates *et al.*, 1990).

These differences implied that the project would need to find ways to control hunting and access more carefully in Okwangwo than in the Oban area, even though vulnerable species occurred in both Divisions, including the forest elephant (*Loxodonta africana cyclotis*), the chimpanzee (*Pan troglodytes*) and the drill (*Mandrillus leucophaeus*).

(f) Gathering

In both areas, a significant source of household income was the collection and sale of non-timber forest products. Some of these, such as rattan canes (*Laccosperma* and *Eremospatha*) and chewing sticks (*Carpolobia*, *Massularia* and *Garcinia*) were harvested destructively (Hurst & Thompson, 1994). Others, such as bush mango fruits (*Irvingia gabonensis*), oil beans (*Poga oleosa*), kola nuts (*Cola acuminata*), and various edible leaves (e.g. *Anthonotha* and *Albizia*), were thought less likely to represent a problem. It was later reported that edible leaves such as those of *Gnetum* were in fact becoming increasingly scarce in the Oban project area (S. C. Heard, personal communication, March 1994).

(g) Montane areas

The Okwangwo forests included an intact sequence from 150 to 1700 m in altitude, which represented a rare opportunity in Africa to preserve a sample of natural ecosystems over such a wide altitudinal range. The upper end of this sequence was on the Obudu Plateau, a unique ecological system with many endemics, which was being destroyed by subsistence farming, uncontrolled fires, and the breakdown of effective management by the parastatal ranching company that controlled most of the Plateau. In the Oban area, by contrast, there was the more usual situation of rugged montane areas being the most remote, least accessible, and least threatened of the Park's ecosystems.

(h) Tourism opportunities

A hotel already existed on the Obudu Plateau which was capable of becoming an excellent tourist facility. This, combined with spectacular scenery, road access, and a pleasant climate, made the Obudu area an especially attractive site for tourism investment. Both the Obudu Plateau and the Mbe Mountains areas of the Okwangwo Division also had potential as sites where rare wildlife such as gorillas might be seen by visitors. The Okwangwo project therefore gave special attention to finding ways to develop this potential.

3.2 Designing responses

3.2.1 Analysis of the task

The purpose of the Cross River National Park Project was to establish the Park legally within appropriate and permanent boundaries. It was therefore necessary to try to resolve all conflicts over land use within the Park. The three main sections of the Park (Oban and Ikpan in the Oban Division, and the Okwangwo Division) each had the potential to contain a biodiversity refuge of several hundred km², which was thought to be large enough to maintain viable populations of most native species. The larger areas to be protected in each location were seen as helping to safeguard overall environmental quality, as well as being able to sustain uses such as traditional harvesting by local people. The total area of the proposed Park, which approached 4000 km², was viewed as the absolute minimum needed for these purposes given the separation of the three sections, the resulting ecological edge effects, and the likelihood of continued encroachment and unsustainable harvesting of resources prior to the full development of the Park.

The rationale for undertaking this project was mainly expressed in terms of the view that forest biodiversity, and environmental services performed by the forests themselves, were major actual or potential assets of the Nigerian and Cross Riverian economies. It was proposed that Nigeria and CRS should invest in protecting natural forests in order to obtain future economic benefits, for example in the form of tourism and research revenues and by avoiding floods and damage to fisheries. Because of uncertainty over the scale of costs and benefits resulting from either preserving or destroying the forest, it was hard to justify the Park only with reference to the national and CRS economies, taken in isolation. External donors were therefore to be asked to subsidize Nigeria's and CRS's investment in the Park. These donors eventually included individual Nigerian and expatriate volunteers, local NGOs (e.g. NCF), international NGOs (e.g. WWF), and governmental agencies (e.g. the British Overseas Development Administration or ODA, and the European Commission or EC and European Union, EU).

All had their own reasons for investing their time and money in subsidizing investment in forest protection. Some took the view that conserving wild species was a sufficient end in itself, but others were also interested in helping the rural communities around the Park to increase their economic well-being. Depending on the donor, the role of short-term rural development was seen either as an end in itself, or as a way to achieve some tangible return on the broader investment, or as a way to help change attitudes among local people, thereby converting them into supporters rather than opponents of the Park.

The size of the area to be protected, the number of people living in and around it or using it, and the intensity of threat to the Park generated by those people, all implied that the eventual project would be large in scale, complex in detail, long in duration, and expensive in cost. It was therefore accepted from the beginning that major institutional donors would be required to sustain the project to completion, and that WWF's role would be to help establish the project and to bring those donors into it. This meant that the start-up and planning phase of the project had to satisfy the needs of two very different groups. The institutional donors had to be persuaded to fund the project, while the Cross Riverians in particular had to support or at least not oppose it. The net result was a project whose design had as much to do with the needs of donor agencies as it had to do with the needs of African people or African biodiversity.

3.2.2 Overview of project planning

The first task was to draw on site surveys by WWF and NCF to prepare a proposal for a start-up and planning operation in the Oban area (Caldecott, 1988f). This proposal resulted in a joint funding agreement with ODA, which allowed WWF to provide management, infrastructure, and logistical support for the planning phase of the project, and to help the CRS Government establish and operate a National Parks and Wildlife Unit. Once this was underway, the next objective was to negotiate an agreement with the EC to undertake a feasibility study for developing the Oban Division of the Park. This was achieved early in 1989, allowing a team of consultants to complete the study in the course of that year. Their reports were then synthesized into a masterplan for the Oban area (Caldecott, Bennett, & Ruitenbeek, 1989). The ODA also commissioned the Natural Resources Institute to undertake a survey of soils, land capability, socioeconomy, and agriculture in the project area (Holland et al., 1989).

The initial planning phase was then completed during the following year by preparing a masterplan for the Okwangwo project area (Caldecott, Oates, & Ruitenbeek, 1990). Both the Oban and Okwangwo feasibility studies included detailed work on park management, fisheries development, forestry, agroforestry, environmental education, and economic appraisal. The masterplans in each case sought to resolve conflicting recommendations by different consultants and to formulate a plan of action and unified budget. Some of the implications of these studies and the practical lessons learned from them were discussed by Caldecott (1991d, 1991e).

The Oban and Okwangwo masterplans concerned themselves with the following main points as they applied to the two Divisions:

- the boundaries, general design and management zones of the Park;
- the structure, function, and objectives of a protection and management service for the Park;
- the structure, function, and objectives of a rural development service for the communities living around the Park;
- the infrastructure and equipment requirements of the project; and
- the need for technical assistance to help the project authorities deploy all resources, given that Nigeria had never before established a rain forest national park.

To obtain the necessary information, consultants were required to document, assess and make recommendations concerning a wide variety of issues (Box 3.2). Recommendations based on these studies included dividing the Park itself into a number of management zones, including:

- **core area** (in Oban) or **conservation and tourism zones** (in Okwangwo), with no hunting or gathering but supervised access for recreation and research;
- **traditional use zones**, for supervised gathering of forest produce by designated local communities;
- **recuperation zones**, reflecting the termination of logging activities and the resettlement of villages; and
- **special use zones**, for example for infrastructure development.

Management plans were produced for both divisions of the Park, specifying a management service which would be responsible for physical protection by providing supervision, law-enforcement and public-relations services as appropriate to each zone. Outside the Park, other zoning categories were used. Certain forest lands were assigned to **special management zones**, where the project would seek ways to change forest management in favour of biodiversity and environmental quality. These were selected where regions were important for endangered wildlife or as watersheds, yet for political or other reasons could not be included within the Park itself. They included the Sankwala Mountains and the Afi River Forest Reserve in the Okwangwo project area.

An assumption of the project was that the greatest long-term threat to the Park was likely to come from the demands of the people living in and around it. The extent to which local people used the forest was therefore assessed, and where significant uses or threats were found, it was proposed to take action through the project to satisfy community needs in ways harmless to the Park. Some 39 villages were found to be partly dependent upon the Park area in Oban, and 66 villages likewise in Okwangwo. These villages were assigned to a **support zone**, where uses of land and environment were to be modified through a **support zone development programme**.

Box 3.2 Subjects investigated during the Oban and Okwangwo feasibility studies.

(a) Soils, terrain and land capabilities (Holland *et al.*, 1989);

(b) current land-use (Holland *et al.*, 1989; Anthonio, 1990);

(c) options to increase crop and livestock productivity in the short and long term (Holland *et al.*, 1989);

(d) socioeconomic patterns in relation to impacts of the park on the people and to traditional land and forest tenure or usage systems, including the role of hunting (Holland *et al.*, 1989; Oates *et al.*, 1990);

(e) environmental education needs (Connor & Martin, 1989; Knight, 1989);

(f) design of appropriate buildings and other facilities (Third & Gibbons, 1989; Ward, 1989);

(g) population and demographic patterns (Holland *et al.*, 1989; Oates *et al.*, 1990);

(h) park management planning for zoning and protection (Schuerholz, Heard, & Sullivan, 1989; Schuerholz, Ojong, & Bisong, 1990);

(i) biological inventories (Reid, 1989; Oates *et al.*, 1990; White, 1990);

(j) constraints and opportunities for tourism development (Caldecott *et al.*, 1990; Erokoro, White, & Caldecott, 1990);

(k) needs, ways, and means for resettling specified communities (Caldecott, Bennett, & Ruitenbeek, 1989; Caldecott, Oates, & Ruitenbeek, 1990);

(l) current agroforestry practices and the potential for agroforestry intervention to help stabilize land-use (Okafor, 1989, 1990);

(m) all aspects of forest management, including the suppression of fire risk and the stable communal use of forest resources, with particular attention to the development of non-timber forest products (Okali, 1989, 1990);

(n) potential for village-level aquaculture development (Taege & Obiekezie, 1989, 1990);

(o) opportunities to link implementation of the project and protection of the Park with relief of Nigerian national debt (Caldecott & Ruitenbeek, 1989; Mundy, 1989); and

(p) economic evaluation of all proposals, and their overall desirability in terms of comprehensive social cost-benefit analysis (Ruitenbeek, 1989, 1990a).

This programme was intended to provide indirect compensation for loss of access to resources in the Park, by improving traditional farming systems and encouraging greater use of agroforestry, by helping communities with natural forests to use them sustainably, and by generally involving local people in the development of the Park. The aim was thus partly to increase wealth among support zone residents by helping them to intensify and improve their use of land. The concept underlying this approach was that large areas of the Park and its support zone were so intimately connected that they should be managed together for a common purpose.

3.2.3 Incentives for local environmental change

The forests of Cross River State were and remain uniquely valuable for conservation, but in 1989–90 they also appeared to be seriously threatened. Based on this perception, the project design emphasized strong measures both to deter further damage to the Park and to provide local people with incentives to conserve. The aim was therefore to achieve maximum impact in minimum time. Protection measures were to be straightforward, involving training, equipping, housing, and deploying staff with which to implement the management plan. Incentives to conserve were more complex, and were mainly to be in the form of benefits available to residents of villages which had registered as part of the support zone, in exchange for their compliance with park management programmes.

These benefits were to include loans from a **revolving credit fund**, which were intended to allow investment in activities judged by the managers of the project to be environmentally benign and economically viable. They were also to include grants from a **village conservation and development fund**, which were to be spent on community works in accordance with the development priorities of each village as agreed with the project managers. In addition, there was to be a **crop loss compensation fund**, which was intended to compensate for damage done around the Park by animals protected within it. As well as these financial measures, certain villages were to be granted exclusive permission to gather forest products (and in the Oban area, to hunt) within traditional use zones of the Park, in exchange for their help in policing those zones and managing them sustainably. The managers of the project would retain the right to suspend some or all benefits in case of serious non-compliance with conservation objectives by people in the support zone.

In designing this package of conditional benefits, it was realized that there was the danger of attracting new settlers to the project area, which would dilute the impact of project resources and increase pressure on lands and forests. This was seen as a serious risk in a country with a population of nearly 90 million, and few controls on internal population movement. Access to all project benefits was therefore to be limited to people who were indigenous to the project area's support zone. The need to define eligibility was avoided by planning to rely on the opinion of the councils of the support zone villages. An indigene was therefore anyone whose claim to be one was endorsed by an appropriate village council. The villages were to be advised that registration of non-indigenes would be against their own interests because it would cause benefits to be diluted.

The feasibility studies concluded that it would be necessary to resettle one village in the Oban Division and three in the Okwangwo Division, comprising about 170 families in total. In addition, relocation assistance was budgeted for another 50 families then living on the Obudu Plateau. Where resettlement was recommended, it was because the villages were deep within the Park and could not be provided with development assistance on site because further growth in settlement size would cause serious damage to the Park's resources.

Resettlement was intended to be voluntary, and the aim was to collaborate with the affected communities in identifying a suitable site to move to, and to work with them in designing and building a new village with a range of facilities not present in the old one. The move was to be accompanied by special access to advice, grants and credit to smooth the transition, and the operation was to be supervised by a full-time resettlement adviser. Tentative proposals to resettle two other communities in Oban and six in Okwangwo were rejected because they would have destroyed distinctive tribal links across the Nigeria–Cameroon frontier, while providing inadequate benefits to the Park to justify the human disturbance and expense involved.

The Mbe Mountains area had a special role in the proposed Park. An NCF gorilla conservation project had been based at Kanyang for some years, managed by Liza Gadsby and Peter Jenkins, and communities around the Mbe Mountains were keen to see some of their forests included in the park so that they could benefit from gorilla-based tourism. A survey team led by Kit Milner was assigned to the area, to work with the local people in selecting, and then marking and mapping boundaries for the Park around the Mbe Mountains. This work resulted in locally approved boundaries in most areas, but these were left out of the National Parks Decree of 1991, which used only existing forest reserve boundaries to define the Park. The latter was probably a matter of administrative convenience. This highlights the need to make it as easy as possible for government practice to absorb new procedures, for example by presenting accurate and legally endorsed maps in good time and in the correct format. Had global positioning system (GPS) and geographic information system (GIS) technology been available at the time, allowing definitive and convincing maps to be presented, the outcome for the Mbe Mountains might well have been different.

3.2.4 Corporate mechanisms

The project design specified that several facilities would be established within the support zone which would generate substantial income for local people, including a livestock breeding centre, a fish hatchery and a central tree nursery (Caldecott, Bennett, & Ruitenbeek, 1989). It was argued that these were required for strategic reasons, to help improve the use of land in the support zone, but they were intended to be self-financing once external subsidy was withdrawn. This implied that they would have to sell their services and make an operating profit. It was therefore proposed to establish a company, Cross River Bioresources Ltd (CRBL), which was to capture profits from new service facilities in the area on behalf of its owners. The latter were to comprise all the inhabitants of all the villages in the support zone of the National Park.

Details of CRBL were to be defined in later studies, but it was modelled on other locally owned corporations which profit from nature reserves and participate in their management, such as those established under the CAMP-FIRE Project in Zimbabwe (McNeely, 1988; Kiss, 1990; Pye-Smith & Feyerabend, 1994). Another factor, though, was the belief that rainforest species within the National Park might yield products with commercial potential. It was thought that CRBL could have a role as the main vehicle for helping such products to be identified, developed, and marketed. This would involve joint-venture research and development programmes with other investors, who would be able to bring specialized expertise, laboratory facilities, and adequate capital to bear on particular projects. Most importantly, the company would ensure that local shareholders would receive a share of the benefits obtained, to reward them for their support in protecting the Park.

3.2.5 Public relations and education

In both the Oban Division (during 1989) and the Okwangwo Division (during 1990), consultants and CRS Government staff spent time at villages, and a side-effect of these visits was to raise local expectations of rapid development of the Park and the support zone. Opposition to conservation later began to be expressed by people who felt that the National Park would exclude them from using the forest in traditional ways. There was a widespread feeling that development had been neglected in the rural areas, and suspicion that the National Park would perpetuate this or even make it worse (S. C. Heard, personal communication, March 1994). These feelings were complicated by a reaction to the high profile of WWF, which was interpreted as foreign rather than local ownership of the project, and also by a sense of injustice over the historical expropriation of forest land for forest reserves and then for the National Park. Confusion over the proposed boundaries of the Park relative to village lands further aggravated the position.

A system to improve communication between the villages and the project was introduced in both project areas during 1989 and 1990, as a result of a suggestion by Francis Sullivan of WWF. Part-time **village liaison assistants** (VLAs) were appointed on the recommendation of their home villages, one man and one woman from each, and were then supposed to continue to reside in those villages. They were provided with special training and participated in village meetings, workshops, and theatrical performances which were intended to contribute to dialogue among the villages, and between them and the project. They were also made responsible for explaining the project to their fellow villagers and for reporting problems and questions back to project staff. This system was seen as a way to achieve a two-way flow of information between the Park authorities and managers of the support zone development programme on the one hand, and the inhabitants of the support zone and the general public of CRS on the other. This would be needed, for example, to advise people about the rules affecting their access to and use of the Park's resources, and to explain the details of project benefits and the terms under which they would be offered.

The VLAs were employed by WWF, and the system developed during 1989–1992 until by April 1992 there were 45 staff representing villages in the Oban area and 25 in Okwangwo (Ashton-Jones, 1992a, 1992b). Although their effectiveness varied among the villages, many VLAs helped to raise awareness of environmental issues in general and of the project in particular (S. C. Heard, personal communication, March 1994). There was an abrupt change in WWF project management in mid-1992, however, leading to a slowing down of all project activities, and a loss of momentum and confidence at all levels in the project including the affected villages (Marshall, 1993a, 1993b). By late 1992, this had caused widespread disillusionment in the project area. These events were partly a result of an unforeseen delay in implementing the recommendations of the feasibility studies. Although a general education message was seen as important in the long term, the delay in funding meant that between early 1990 and early 1994 in the Oban area, the VLAs had little of practical urgency to communicate other than that the project was 'about to start'. This eventually cost the project much local credibility, especially after mid-1992.

Awareness of the aims and benefits of the project thus probably peaked too soon, relative to the beginning of full-scale implementation. Moreover, rather than using the trained human resources of the communication programme when the project began to be implemented in full, the VLAs were allowed to disperse. There were several reasons for this, including WWF's need to disengage itself from a large staff in view of the uncertainty over its role in the Oban project area after 1993 (R. Barnwell, personal communication, August 1994). An overall conclusion is that intensive communication and education efforts can work well at the village level, but need to be paced relative to project resources and co-ordinated closely with other project activities.

3.2.6 Responding to other needs

In 1989–90, commercial forestry operations in CRS were almost paralysed by equipment failure and a shortage of capital. There remained the possibility, however, that strong national demand for wood and wood products would allow them to recover sufficiently to threaten the Park, and efforts were in fact underway to refinance at least one logging company. It was clear from early in the design phase of the project that creating the Park would mean that more than half of the State's forest land would no longer be available for timber production. This was thought to be a possible source of opposition to the project, and ODA therefore sent a team in early 1990 to assess the need for forestry sector assistance. The aim of this was to help the State adjust to the creation of the Park, by encouraging more efficient and sustainable use of its remaining production forest.

As a result, an ODA technical assistance programme began its first three-year phase in late 1990, aiming to advise the Forestry Department on natural and plantation forest management, forest product use, and on the equipment and training needs of the Department itself. Meanwhile, the Ashland Oil Company provided a helicopter with which to survey forest areas outside the Park. This allowed an area of about 250 km^2 south of the Oban Division to be identified as a suitable site for a sustainable wood and rattan harvesting project. During its first phase, the forestry project also promoted new ideas about sustainable forest management and community participation within the Forestry Department, and by 1993 had made considerable progress in doing so (S. C. Heard, personal communication, March 1994). This success had, however, resulted in some frustration among staff of the Forestry Department since few resources were available with which to put the new ideas into practice. An exception was the Ekuri Community Forestry Project, which resulted from collaboration between two villages and the forestry and park projects, and demonstrated the value of strengthening community ownership and responsibility for managing natural resources (Box 3.3).

Box 3.3 The Ekuri Community Forestry Project
(S. C. Heard, personal communication, 1994)

Old and New Ekuri are two isolated villages in the north-western part of the support zone of the Oban Division of Cross River National Park. They share about 250 km² of community-owned forest land adjacent to the Park, and since the mid-1980s had been exploring ways to use this forest for maximum communal benefit. They became involved in discussions with National Park personnel about the possibility of managing their forest for sustainable yields of timber and non-timber forest products, and in 1992 an expatriate volunteer Community Forest Officer (CFO) was assigned to the area. Within one year, the communities had completed an inventory of a 0.5 km² plot and produced a stock map, while training 16 people in demarcation and enumeration skills, and reaching a position of being able to determine which trees should be harvested by the two communities. Just as important, a previously antagonistic relationship between the villages and the Forestry Department had become much more positive.

This progress was attributed to a number of factors, including: the role of the CFO in promoting intensive discussion about forest management at the village level; local confidence in the CFO prompted by his residence in the villages, and by his actively helping the villages to obtain new bridges to improve access to the area; the active role of the CFO in articulating requests for assistance and promoting collaboration with the Forestry Department; and both the involvement of women in the forest inventory and promotion of a sense of community ownership of the project. The Ekuri Community Forestry Project showed that it is possible to achieve community participation in forest management quickly and cheaply, if sufficiently skilled and responsive help is provided. This shows how talented and motivated volunteers can catalyze community initiatives, but simultaneously raises the question of how to achieve unassisted, self-starting programmes of the same kind within traditional rural societies.

Because the National Park project envisioned considerable expenditure within Cross River State, there was strong support for it within the CRS Government, and little anticipated opposition outside the forestry sector. At the Federal Government level there was consensus in favour of creating Cross River National Park, but not necessarily the willingness to assign a large part of Nigeria's foreign aid receipts to pay for it. In this sense, the Park was in competition with other projects elsewhere in the country, in places which might be expected to yield greater political advantage nationally than by helping Cross River State. This was relieved by using EU regional funds, which could only be obtained in the context of co-operation with Cameroon.

The project was also formulated in such a way as to emphasize its benefits for Nigeria as a whole. Of special importance in creating the willingness to invest at the Federal level was the inclusion of specific means to link the long-term protection of the Park with relief on some of the obligations created by Nigeria's large external debt. These proposals focussed especially on intergovernmental or sovereign indebtedness, and resulted in WWF being authorized by the Federal Ministry of Finance and Economic Development to explore the establishment of a management fund comprising sovereign debt instruments of up to US$300 million 'for the purpose of defraying the recurrent cost of Cross River National Park' (FGN, 1989:1). This interest was presumably helpful to the National Park project, since the same ministry was responsible for authorizing the request for EU funding with which to implement it.

3.3 Discussion

The Cross River National Park (CRNP) project was designed in response to the divergent needs, demands and habits of thought of conservationists, development agencies, governments and local farmers. This happened at a time when WWF and the European Commission, like other NGOs and governmental donors, were beginning to learn how to work together on large, complex conservation and development projects. The project was also WWF's first major field operation in Nigeria, and it was Nigeria's first attempt to create a rain forest national park. There was thus a lack of specific and relevant experience among all the groups and most of the individuals involved in planning the project. The group with most relevant experience in fact probably comprised the local people in the project area, who had been many times exposed to government plans and projects affecting the forests and lands which they regarded as their own. This was not fully reflected in either the consultation or the planning process that defined the project.

The project was designed on the assumption that the forests within the Park were under immediate threat by local people, and this assumption was never properly justified. It was an impression gained mainly from short-term field work in the project area, and from the general state of Nigeria's forests outside CRS. It was supported by dramatic events such as the burning of scrub and some forest overlooking the settlement of Oban at the beginning of 1989 (just prior to a visit there by HRH Prince Philip, the International President of WWF), by fires in the forest fringes of the Mbe Mountains, and by the abundance of rare wildlife species as bushmeat in the rural markets. What it lacked was an understanding of the context of these observations in the long-term social and economic life of the project area. There was at least the possibility, therefore, that the intensity of threat to the Park area was overestimated at the beginning of the project.

The assumption of extreme threat affected the recommendations of the design team, especially those concerning the need for resettlement, and for the mixture of enforcement measures and benefits to reward compliance with them. Since the Park enforcement measures were straightforward to implement, whereas the benefit packages were inherently complex and expensive, the result was that the former overtook the latter. This had an inevitable impact on public relations in the project area, which was only partly compensated by the education and community awareness programme described above.

An important factor was that the public in the project area had no previous experience of the enforcement of conservation rules, and friction arose when this was attempted (U. E. Ité, personal communication, July 1994; Ité, 1995). A review of the Okwangwo project in July 1994 concluded that the Park was indeed acutely threatened, and reaffirmed the need for all the measures specified in the 1990 masterplan (Hurst & Thompson, 1994). The report also recommended additional efforts to protect the Park against exploitation and encroachment, combined with vigorous public relations measures, and strongly advocated resettlement of communities within the Park as originally planned.

One source of public hostility encountered by the project was opposition to the thesis that a government has the right to grant and withhold privileges to use resources in forest areas which had already been so used for generations. This issue is an important and complex one, which affects many conservation projects. One way to resolve it is to involve local people much more closely in the decisions which generate such conflicts than could be achieved in the design phase of the CRNP project. A constraint on doing this was the schedule of the feasibility studies in 1989–90. These were carried out by WWF under contract to the European Commission, and it was then perceived that the overwhelming priority was to complete them to a high technical standard, on time and on budget.

These aims were achieved, but at the cost of reliance on short-term consultants. This was partly compensated by using several Nigerian experts and others with prior long-term experience in Nigeria. The consulting team was never in a position, however, to take the time to listen at length to local people or to help them articulate their own priorities for development. Its job instead, in a few months, was to produce a set of technical reports, a masterplan and a plausible budget, which could be used to activate Nigerian Government procedures to constitute the Park, and EC procedures to allocate the financing for it.

There followed a long delay between finishing the Oban masterplan and implementing those of its recommendations which were intended to benefit the local people directly. This caused a loss of credibility, and made the project vulnerable, for example, to rumours that funding had in fact been provided but had been stolen before reaching the project area (U. E. Ité, personal communication, July 1994). The masterplan was completed and submitted for review in August 1989, before being finalized in November 1989 and accepted by the Federal Government in December 1989. At that time, the Federal Government asked the European Union for assistance with which to implement the project. In April 1990, a draft financing proposal was prepared by the EC Delegation in Lagos, and was subsequently approved by the Commission in Brussels.

A team from the Commission and the German Development Credit Agency (*Kreditanstalt für Wiederaufbau*) visited Cross River State in June 1991 to assess the feasibility of the project and to amend its structure in preparation for putting it out to international tender. Consulting firms were short-listed for the Oban Hills project management contract in March 1993, and the contract was awarded in October 1993 to a consortium led by the British consulting firm Hunting Technical Services Ltd. This meant that a period of more than five years had elapsed between the beginning of project planning and the beginning of project implementation. Even though, perhaps, threats were not as extreme in this particular project area as at first thought, such a schedule would be quite inadequate to resolve many of the serious conservation problems that do arise regularly worldwide. If major donor agencies wish to contribute to conservation, therefore, they must find ways to act and react more quickly than the European Commission and its partners were able to do in the case of the Oban project.

Had WWF been better prepared for this slow pace of activity, it would have been possible to avoid raising expectations in the project area. A programme might instead have been undertaken involving basic conservation work around the obvious priorities. These were and remain:

- to protect the forest;
- to maintain lines of communication between the people affected by that protection and the people doing the protecting; and
- to help both sides understand the ecological limits of their environment and how to live the best possible lives without exceeding those limits.

A final observation is that, while the project gradually unfolded, the Nigerian Government proceeded to fulfil its stated intention of creating the Park. It achieved this despite its own political and economic difficulties, which included attempted *coups d'état*, factional riots, general strikes, and financial crises on an enormous scale. Despite these interruptions and precisely on schedule in October 1989, the Federal Nigerian Council of Ministers approved Cross River National Park to include both the Oban and the Okwangwo areas. This led in due course to Presidential Decree No. 36 of 1991, which legally created Cross River National Park on the 2nd October 1991.

4

Siberut and Flores Islands, Indonesia

4.1 Conservation issues

4.1.1 Land, sea, and people

Indonesia embraces about 17 000 islands with a land area of nearly 2 million km^2 spread over 5100 km between mainland Asia and Australia (RePPProT, 1990). They have some 81 000 km of coastline among them, or 14% of the world's total, and are set in 3 million km^2 of territorial sea which links the Indian Ocean to the Pacific (Whitten & Whitten, 1992; Fig. 4.1). Western Indonesia includes lands rising from the Asian (Sunda) continental shelf, including the islands of Sumatra (475 000 km^2), Java (133 000 km^2) and Borneo (740 000 km^2), of which 536 000 km^2 is in Indonesian Kalimantan. Eastern Indonesia includes lands rising from the Australian (Sahul) continental shelf, including the islands of Kai, Aru, and New Guinea (867 000 km^2), of which 415 000 km^2 is in the Indonesian territory of Irian Jaya (Chapter 7). These shelves are covered by seas often less than 200 m deep, but between them depths may exceed 8000 m. The central islands are thus separate from both continental shelves, and include Sulawesi (186 000 km^2), Nusa Tenggara (81 000 km^2, including East Timor), and Maluku (78 000 km^2).

Indonesia is the fourth most populous nation in the world. In 1993, its total population was about 188 million and each year was growing at about 1.7% and expected to increase by some 90 million within 35 years (Dompka, 1994). Nearly two-thirds of all Indonesians live on Java and on the nearby islands of Madura and Bali. This is because of their long history of advanced civilization, supported by irrigated farming on their fertile volcanic soils. The Indonesian people are culturally diverse, with several hundred distinct ethnolinguistic groups. The Indonesian language is used for formal purposes throughout the country. It arose from Malay, a trading language used

61

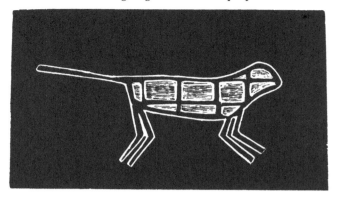

Siberut monkey-tatoo design.

throughout the Malay Archipelago (which includes Malaysia and the Philippines), and formerly as far away as Madagascar.

4.1.2 Biogeography and biodiversity

The shallow modern seas over the Sunda and Sahul Shelves have come and gone, as sea level changed over the last 2 million years in response to successive global Ice Ages and warming events. This factor exerted a strong influence on Indonesia's biogeography, since at times it promoted the dispersion of terrestrial species through the Sunda and Sahul land masses while limiting dispersion between them, and at other times it tended to promote isolation and local speciation within the various islands (BAPPENAS, 1991; KLH, 1992). These influences are seen in the division of Indonesia between the Indo-Malayan and the Australasian zoogeographical realms. Animals native to the former are mostly related to Asian species (e.g. the placental mammals), and those in the latter mostly to Australian ones (e.g. the marsupial mammals). There is a transitional zone between them ('Wallacea'), where animals can be related in either direction (Wallace, 1869). Among plants, groups centered on the Sunda Shelf include the dipterocarps and rattans (Chapter 2), and the Magnoliaceae and *Artocarpus*, while those on the Sahul Shelf include the Araucariaceae, Winteraceae, and *Styphelia* (Whitmore, 1989).

A distinct and very diverse phytogeographical region known as 'Malesia' extends from the Malay Peninsula to the Philippines and New Guinea, and therefore includes all of Indonesia (van Steenis, 1950; Whitmore, 1984).

Regional differences in the flora mean that Indonesia includes the South Malesian (Java, Bali, Madura, and Nusa Tenggara) and parts of the West and East Malesian Botanical Sub-Regions. The country's seven major land masses and island groups thus all have distinctive arrays of local ecosystems and wild species, the details of which are still far from fully documented (Whitten *et al.*, 1984; Whitten, Mustafa, & Henderson, 1987; Whitten, Soeriaatmadja, & Afif, in press; MacKinnon *et al.*, in press; Monk, de Fretes, & Lilley, in press).

The whole of Indonesia lies within a few degrees north or south of the equator, and the larger land masses of Sumatra, Borneo, and New Guinea have moist equatorial climates with a mean annual rainfall of 2500–5000 mm. The middle of the country is strongly influenced by the dry southern monsoon which blows off Australia in May to September (Dick, 1991). This creates a broad band of seasonal and semi-arid climates in most of Java, Nusa Tenggara, Maluku, and Sulawesi, with mean annual rainfall in parts of Nusa Tenggara as low as 700 mm (MoF & FAO, 1991). The range of climates, other sources of spatial variability such as topography, soils, island size, and geological age, as well as biogeographic influences, all combine to give Indonesia a wide range of natural ecosystems.

These support at least 20 000 species of flowering plants, 515 mammals, 900 amphibians, and reptiles, 1520 birds and 4200 fish, an overall total of 15% or more of all the species in the world (BAPPENAS, 1991; KLH, 1992; WCMC, 1992, 1994; Whitten & Whitten, 1994). Although most invertebrate taxa are poorly studied, the dragonflies (666 species) and swallow-tail butterflies (122 species) are both known to be more numerous than in any other country (WCMC, 1992). Available data indicate that species richness varies greatly among the major island groups (Table 4.1).

Indonesia is located in the middle of the Indo-Pacific Region, a marine biogeographical division stretching from the Red Sea to Polynesia. This Region possesses more than 4000 non-pelagic fish in 179 families, and has the richest shorefish fauna in the world (Myers, 1991). Almost all of these families and at least half the species occur in Indonesian waters, which are probably the most species-rich in the world (Norse, 1991). Many of these fish are associated with coral reefs, and Indonesia possesses at least 500 species of corals and thousands of species of invertebrates associated with reef environments. Indonesia's coral reefs are the most species-rich of any in the world (UNEP & IUCN, 1988; Kuiter, 1992; Veron, 1993; Lieske & Myers, 1994; Tomascik *et al.*, in press).

Many of the Indonesian islands have been isolated for long periods from one another, and have evolved local endemic species. The endemism rate

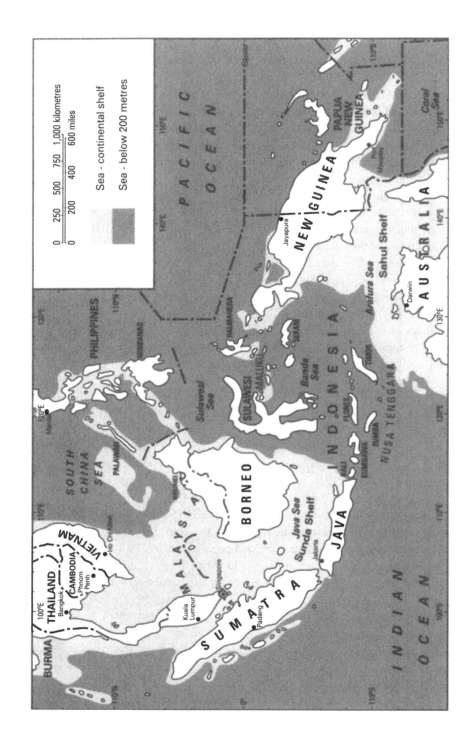

Table 4.1 *Species-richness by taxon and Indonesian island group (from BAPPENAS, 1991; Kottelat et al., 1993; Chapter 7)*

Island group	Number of native species: Birds	Mammals	Reptiles	FW fish	Selected plant taxa
Sumatra	465	194	217	272	820
Java	362	133	173	158	630
Kalimantan	420	210	254	368	†900
Sulawesi	289	114	117	68	520
Nusa Tenggara	242	41	77	–	150
Maluku	210	69	98	–	380
Irian Jaya	639	154	223	316	1030

Table 4.2 *Species endemism by Indonesian island group (from BAPPENAS, 1991; Kottelat et al., 1993)*

Island group	Per cent of endemic species: Birds	Mammals	Reptiles	FW fish	Selected plant taxa
Sumatra	2	10	11	11	11
Java	7	12	8	16	5
Kalimantan	6	48	24	38	33
Sulawesi	32	60	26	77	7
Nusa Tenggara	30	12	22	–	3
Maluku	33	17	18	–	6
Irian Jaya	52	58	35	–	55

varies between island groups (Table 4.2), and for Indonesia as a whole it ranges from 23% among birds, to 26% in fresh-water fish (western Indonesia only), 32% in mammals, 37% in amphibians, 43% in swallow-tail butterflies, 48% in dragonflies, 59% in reptiles, and 67% in higher plants (WCMC, 1992; Kottelat *et al.*, 1993). Especially high rates of endemism are encountered in Sulawesi, in Irian Jaya and (among non-flying mammals) in the Mentawai islands off the west coast of Sumatra (see below).

Fig. 4.1 Indonesia

Designing conservation projects

Table 4.3 *Indonesian forests in 1990 (from KLH, 1992) – areas of*
production forest (PF) and unclassified forests (UF) in km²

Region	PF un-logged	Lightly logged	Heavily logged	UF	All forests
Sumatra	56 700	18 100	9 000	36 000	119 700
Kalimantan	107 900	40 300	19 200	59 700	227 700
Sulawesi	26 300	1 900	8 000	15 300	49 500
Maluku	17 300	2 200	1 100	19 600	40 200
Irian Jaya	95 500	1 000	300	60 900	157 700
Nusa Tenggara	800	100	800	3 500	5 200
TOTAL	304 500	63 600	38 400	195 000	600 000

4.1.3. The timber industry

Commercial logging began in the early 1960s in Kalimantan, and within 10
years, large-scale logging was widespread there and elsewhere in Indonesia
(Nectoux & Kuroda, 1989). Concessions had been awarded by 1988 covering
about 434 000 km² or three-quarters of the total Indonesian forest estate
(Collins, Sayer, & Whitmore, 1991; Table 4.3). The Indonesian Government
sought to phase out the export of logs from the mid-1970s, a policy linked
to its aim of becoming the world's largest producer of plywood. In 1981,
Indonesia exported 7.2 million m³ in the form of logs and 0.8 million m³ as
plywood, but rapid growth in the plywood industry meant that by 1991 these
figures had been reversed, with exports of only 1.4 million m³ as logs and
8.2 million m³ as plywood (FAO, 1993). The total log harvest, meanwhile,
rose from about 16 million m³ in 1981 to about 26 million m³ in 1987 (Gillis,
1988b), and the annual target harvest for 1995–2000 was reported to be over
37 million m³ (Anonymous, 1994).

The development of Indonesia's timber industry was based on a consensus
classification of forest function (TGHK, *Tata Guna Hutan Kesepakatan*).
This arose from 1970 to 1985, and involved discussions among various
government agencies to produce maps showing agreed allocations of forest
land to various kinds of permanent use category (Dick, 1991). The five main
TGHK categories and their forestry uses were as follows:

- *Nature Reserve* (no timber extraction);
- *Protection Forest* (no timber extraction);
- *Limited Production Forest* (non-industrial selection felling);

- *Regular Production Forest* (industrial selection or clear felling according to forest type); and
- *Conversion Forest* (clear felling and conversion to other uses).

The TGHK mapping programme was flawed mainly by a lack of adequate information to support spatial and forestry planning (Dick, 1991), but other mapping operations were undertaken to correct this. The most comprehensive was by the Ministry of Transmigration in the late 1980s, which mapped land-use and land capability for the whole country outside Java and Bali, with a view to finding suitable places to receive officially sponsored settlers (RePPProT, 1990). The Ministry of Public Works has since integrated the RePPProT maps on a common scale with TGHK and District and Provincial planning maps, and has updated them from field observations and new remote imagery. These maps now show actual forest cover, Protected Forests, Nature Conservation Areas, Sanctuary Reserves, and the alignments of existing and proposed roads and other development projects (C. Vandersluys, personal communication, September, 1994). A comprehensive legislative context for a national system of spatial planning was provided by Act No. 24 of 1992, although cases of conflict between planned and actual uses of land continue to occur (Dahuri, 1994; World Bank, 1995; Chapter 7).

4.1.4 Environmental threats

In Indonesia, high levels of ecosystem diversity, species-richness, and endemism exist alongside a rapidly expanding economy and a large, growing, and internally mobile human population, much of which is rural and engaged in agriculture. Dick (1991) summarized the process of deforestation in Indonesia, discussing the relative impacts of commercial forestry operations, development of tree crop estates, swampland conversion, the role of smallholders, forest fires, and transmigration programmes. The extent to which natural forests have been cleared approached 44% for the country as a whole by the early 1980s, although the extent of remaining forest varied by province from 6 to 11% in Java, 19 to 69% in Sumatra, 46 to 72% in Sulawesi, to 84% in Irian Jaya.

Indonesia is such a large and diverse country that many different threats are found there, often in combination with one another. Habitat loss has been most acute on the densely populated islands of Java and Bali, where the fertile lowlands have long been farmed and remaining natural forests are mainly montane relics or lowland patches on poor soils (Collins, Sayer, &

Whitmore, 1991). Threats elsewhere include the widespread conversion of natural forest by mines, plantations, small-scale encroachment, and trans-migration schemes, and the degradation of wetland and coastal ecosystems by industry, agriculture, sedimentation, mining, destructive fishing methods, and by drainage and the building of aquaculture ponds (World Bank, 1990; BAPPENAS, 1991; Hardjono, 1991). Replacement of traditional resource tenure systems, direct over-harvesting of wildlife populations, and the eco-logical effects of introduced species are also widespread problems throughout Indonesia.

These factors vary greatly in the extent to which they are encouraged or can be discouraged by factors acting at the different levels of Indonesian society. In many locations in Sumatra, Sulawesi, and Nusa Tenggara, for example, the major immediate threat to natural environments is local encroachment by growing populations in need of farmland and fuelwood. In Irian Jaya and Kalimantan, by contrast, the more important threats tend to be from logging, mining, road building, and officially sponsored transmigration. These two kinds of threats pose quite different challenges. The first is driven by local people whose day-to-day actions at the family level degrade environ-ments gradually at the extensive forest margin. The second, however, is driven by large institutions or industries, which create much more specific and immediate demands for access to large units of resources.

Some government policies have an important role in aggravating threats to the environment, especially those which maintain artificially low timber prices, encourage short-term, single-output forest management systems, or which subsidize the conversion of natural forests to monocultures of exotic trees (BAPPENAS, 1991). Policies were also developed in the early 1990s to encourage pulp and paper production in Indonesia (DtE, 1991). These are expected to increase the clear-felling of large areas of accessible mangroves and logged-over forests, and the replacement of the latter with plantations of fast-growing pulpwood.

4.1.5 Measures to protect biodiversity

Planning for a national system of protected areas in Indonesia occurred in parallel with TGHK, and later with RePPProt. By 1990, Indonesia had gazetted 303 terrestrial nature reserves of various kinds totalling 160 000 km^2 or 8.2% of land area, and another 20 000 km^2 in 175 sites had been proposed as such reserves (MoF & FAO, 1991). These areas were chosen so as to include viable and representative samples of most ecosystems, and populations of most native species (MacKinnon & Artha, 1981–2). More than 300 000 km² had meanwhile been designated through TGHK and RePPProT as protection forest, with the main role of safeguarding water catchments and steep slopes (BAPPENAS, 1991). A total of 23 marine and coastal areas had also been protected, with another 200 formally proposed, and the official intention was to have about 300 000 km² of marine reserves legally designated by the end of the century.

The terrestrial reserves included a wide range of ecosystems on most of the major islands, and a majority of native birds and conspicuous mammals had been recorded from at least one designated reserve. Some gaps in coverage remained to be corrected, however, for example the endemic bird areas of Maluku as mapped by ICBP (1992). Once completed, the reserve system is expected to give good coverage on paper for most components of Indonesia's biodiversity. Up to the mid-1990s, however, many reserves had little effective management, reflecting the limited resources of the Directorate General of Forest Protection and Nature Conservation (PHPA, part of the Ministry of Forestry).

Indonesia's conservation efforts have been supported by international donors, and levels of aid for this purpose reached about US$12 millions annually in the early 1990s (KLH, 1992). At that time, rates of expenditure by PHPA and its partners averaged about US$75 km⁻² yr⁻¹ for the reserve system as a whole (KLH, 1992). This can be compared with a recommended minimum figure of US$300–400 km⁻² yr⁻¹ for national park management suggested by IUCN (1992), and levels of expenditure 300–500% higher than Indonesia's for priority nature reserves in Thailand and China (KLH, 1992). In any case, field reports from Indonesia clearly suggested that conservation investment was often insufficient or misdirected (McCarthy, 1991). To bring average expenditure rates in Indonesia into the same general range as Thailand's and China's, at least US$130 million per year would be required (KLH, 1992).

Levels of assistance for Indonesia's system of nature reserves were expected to increase during the 1990s as donors committed themselves to fund projects identified under the national *Tropical Forestry Action Programme* (MoF & FAO, 1991) and *Biodiversity Action Plan* (BAPPENAS, 1991). An important model for such work was that of 'integrated conservation-development projects' (ICDPs), which seemed to offer a more holistic approach than others to providing development among rural people, and especially those living around nature reserves. Although there was later an emphasis on less expensive forms of conservation action in Indonesia (Chapter 7), the ICDP concept in the early 1990s was an attractive one which

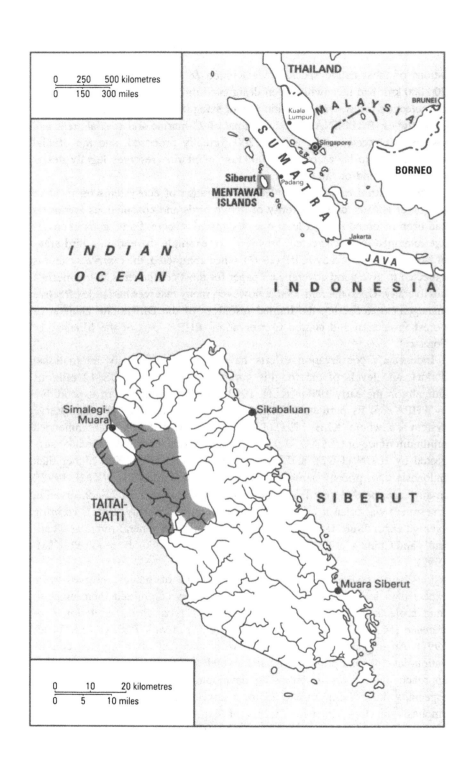

was actively explored by government and by institutional donors and international NGOs working in Indonesia.

An example is provided by the collaboration between PHPA and the Asian Development Bank (ADB) in designing and implementing a *Biodiversity Conservation Project in Indonesia*. This aimed to strengthen institutions and human resources at the national level, while simultaneously implementing field operations in two selected critical areas. It proceeded from a review of Indonesian conservation areas and selection of potential sites, which ADB commissioned from Deutsche Forst-Consult (ADB, 1992a), through a pre-appraisal by the Bank (ADB, 1992b), and subsequent appraisal, negotiation, and award of the main management contract. The two sites selected for field work were Siberut island off the cost of Sumatra, and the Ruteng area of Flores Island in Nusa Tenggara.

4.1.6 The island of Siberut

With an area of about 4480 km², Siberut is the largest of the Mentawai Islands, and lies about 130 km off the west coast of Sumatra within the Province of West Sumatra (Fig. 4.2). The Mentawai Islands have been separated from mainland Sumatra by the 1500 m deep Mentawai Strait for at least 500 000 years (Whitten, 1980, 1982; Avé & Sunito, 1990; McCarthy, 1991). This isolation has resulted in the Mentawai islands and especially Siberut having a higher density of endemic non-flying mammal species per unit area than any other part of Indonesia, and high levels of endemism also in other taxa. Up to 15% of the flora, 13% of the birds, and 65% of all mammals are endemic at the species or subspecies level. Endemic mammals include 5 squirrels (*Callosciurus, Lariscus, Sundasciurus, Iomys,* and *Hylopetes* spp.), 3 rats, a palm civet (*Paradoxurus lignicolor*), and all 4 native primates (a gibbon, *Hylobates klossii*, a leaf monkey, *Presbytis potenziani*, a macaque, *Macaca pagensis*, and an odd-nosed monkey, *Simias concolor*).

The geology of Siberut is mainly sedimentary, and the island is rugged and deeply dissected by rivers and streams, fed by a mean annual rainfall of over 3300 mm with little seasonality. Soils are heavily leached and very poor in nutrients, with high erosion potential leading to the deposition of silt in lower reaches of rivers and the accretion there of broad alluvial terraces

Fig. 4.2 The Mentawai Islands

(WWF, 1980; Mitchell, 1982). Tectonic movements have created a subsiding, irregular east coast, and earth tremors are frequent in the area. Dipterocarp forests dominate hilly areas of Siberut up to the highest point at 384 m. Other forest types are fresh-water swamp forest, commonly with abundant sago palms (*Metroxylon sagu*), mangrove forest along the leeward east coast, and *Barringtonia* forest along the windward west coast. The dipterocarp forests have attracted the attention of the timber industry, and most of the island was allocated to logging concessions in 1973. Logs were extracted at a rate of about 150 000 m³ yr⁻¹ during the 1970s (WWF, 1980) and logging continued into the 1990s. Large quantities of rattan canes and mangrove poles were also extracted from Siberut over the same period.

The people of Siberut are culturally distinctive, having arrived on the island several thousand years ago with a Neolithic culture and thereafter having been isolated from events elsewhere (WWF, 1980). There was thus no pre-modern contact with Buddhist, Hindu, or Muslim cultures which successively dominated the rest of the Malay Archipelago. A rich ethnobiological tradition developed on Siberut due to the inhabitants' dependence on the living resources of the island (Avé & Sunito, 1990). The aboriginal population is also divided into about fifteen different linguistic groups, which occupy separate river systems.

Contact between the Mentawaian population and modern Indonesia was traumatic, especially during the 1950s and the 1970s when government policy involved resettling people from communal clan houses into new villages, introducing rice-based cropping systems at the expense of sago-based ones, and suppressing local traditions, either directly or through religious conversion (Avé & Sunito, 1990). This external pressure eased during the 1960s, but the 1970s saw the award of logging concessions and the beginning of large-scale logging of the island. There were serious environmental and social impacts, as this logging occurred on steep slopes with fragile soils under high rainfall, within forests used for hunting and gathering by local people.

Social pressures eased again during the 1980s, but immigration by non-aboriginal people, introduction of inappropriate cash crops, food crops and livestock, and exposure of Siberut to market forces continued to disrupt local life. One feature has been the overharvesting of rattans (*Calamus* and *Daemonorops*) and gaharu (perfumed wood from certain *Aquilaria* trees – see Chapter 2). In 1990, plans were revealed which, if implemented, would have converted over half of the land area of Siberut into oil palm plantations (Tenaza, 1990).

The catalogue of profound disturbance on Siberut is thus impressive, and although attempts were made to plan the island's development in a sustainable manner (WWF, 1980; Mitchell, 1982), these were overwhelmed by opposing interests. In 1992, however, central government announced its intention to phase out logging on the island by not renewing concessions (which prompted a sudden acceleration of logging activity). At about the same time, decisions were taken to suspend resettlement and agricultural conversion, and to create a large nature reserve on Siberut, while also placing the island's development under the guidance of a multi-sectoral committee of concerned agencies at the Provincial Government level. These decisions followed a prolonged national and international NGO campaign on behalf of Siberut's environment and people. They also coincided with interest by the ADB in making a large grant and soft loan to Indonesia with which to conserve the island's biodiversity and develop its economy sustainably.

4.1.7 The island of Flores

The forested mountain range south of Ruteng, capital of Manggarai Regency in western Flores (Fig. 4.3), is in the wetter part of Flores island, but still receives little more than 1000 mm of rain annually, most of which falls during the monsoon from November to April. The forest is a mixture of lowland monsoon and montane rainforest formations and is of great importance for biodiversity conservation since it is the largest (indeed, almost the only significant) surviving moist-forest fragment in Nusa Tenggara (Collins, Sayer, & Whitmore, 1991). It thus has a disproportionate role in maintaining species richness within one of Indonesia's biogeographical regions. The flora and fauna of the area are poorly known (Monk, de Fretes, & Lilley, in press), but from its location it is likely to be an important refuge for a number of restricted-range birds (ICBP, 1992, 1995).

The Ruteng forests have direct economic significance since they protect the watershed which sustains Ruteng town and agriculture in half the Ruteng valley to the north, and down to the coast southwards. The considerable tourism potential possessed by the Ruteng area is also dependent on maintaining natural vegetation on the slopes surrounding the town. In 1992, the forest area comprised a series of contiguous Protection Forests and Limited Production Forests, all to some extent degraded by long-term, low-technology extraction of firewood and construction timber. There was also a history of encroachment by terraced and other farming, although the established boundaries of the forests were in general respected by the local people. Serious poverty might have been a factor close to the reserve, since Nusa Tenggara is one of the poorest regions of Indonesia. In 1986, it had an average GDP

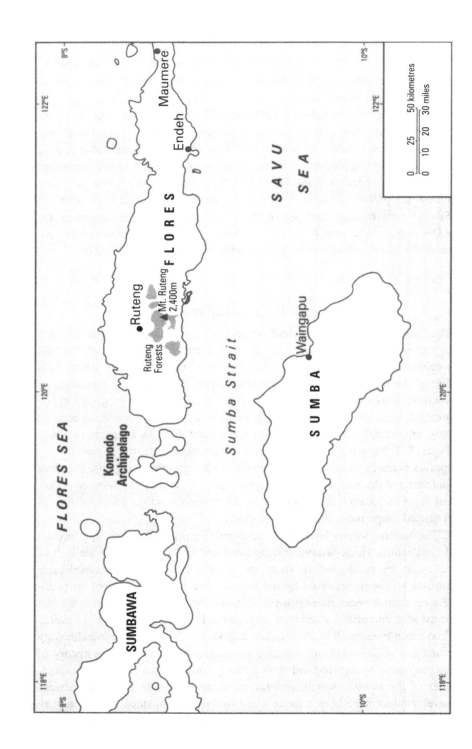

per person only about 20% that of Jakarta and 8% that of East Kalimantan (MoF & FAO, 1991).

The seasonal climate makes the area vulnerable to fire, and although some fire damage had occurred, increased sedentary agriculture and the establishment of artificial forests in the surrounding area seemed to have reduced this danger somewhat. There are also geological hazards to the forest, since the Inner Volcanic Arc of Indonesia runs directly through the area. A volcanic eruption occurred in 1990 near the crest of the Ruteng mountains, and a crater and large cinder slope were still visible from the air approaches to Ruteng town in 1994. The area was little affected, however, by a major earthquake in Flores in December 1992, which devastated the provincial capital of Maumere about 160 kilometres to the east.

4.2 Designing responses

4.2.1 The Siberut project

In the Foreword to WWF's (1980) masterplan for Siberut, the Indonesian Minister of State for Development Supervision and the Environment, Dr Emil Salim, observed that: 'Siberut ... plays a unique role in Indonesia's natural and cultural heritage. The endemic plants and animals are so special that there is simply no reasonable alternative to taking all possible steps to protect them. At the same time, the unique inhabitants of Siberut are "isolated people" whose well-being is of outstanding interest to the country. This plan shows how the values of the Siberut culture can be maintained while it is gracefully brought into the mainstream of Indonesian society and the natural environment is saved as well.'

The prescriptions of this 1980 masterplan included:

- improving welfare by increasing the efficiency of the subsistence economy;
- zoning the island for various kinds of land use;
- creating a protected area on Siberut;
- constituting the whole island as a biosphere reserve;
- improving forestry practice so as to minimize logging damage;

Fig. 4.3 Western Flores and Ruteng

- encouraging culturally and environmentally benign forms of tourism;
- managing wildlife while maintaining traditional hunting;
- educating Siberut people about sustainable land-use, and educating other Indonesians about the values of Siberut; and
- using local and national committees to evaluate and monitor progress.

In 1992 as in 1980, the main threat to biodiversity on Siberut was thought to be the ongoing logging in many parts of the island. In 1992, however, most of the island's timber resources had been felled, logging was in its final stages, and the ADB project was conditional on logging being ended. Other plans or attempts to implement large-scale plantations and transmigration projects on Siberut had been abandoned, but were thought to have left a degree of environmental damage which the project would need to help repair. Residual threats to biodiversity included continued forest clearance for agriculture to support a growing population and demand for trade goods, erosion of damaged lands, hunting using new weapons, and the potential risks of excessive harvesting of certain species of wild plant and animal for trade and subsistence use.

Because of the history of environmental usage on Siberut, the first stage of the ADB-financed project was intended to be a thorough survey of the island to assess impacts to date, and to define the need for urgent restoration work. The aims of the Siberut project thereafter are summarized in Box 4.1. Its design started with the assumption that, because of poor soils and drainage, difficult terrain and high rainfall, it is unlikely that Siberut will ever be a sustained source of great wealth using orthodox development procedures. The project was not intended to achieve this, and instead the aim was to stabilize conditions and to encourage steady improvements through maximum local capture of such wealth as could be created through benign use of the environment.

At the same time, these steps would preserve the option and lay the groundwork for Siberut and Indonesia to benefit from the potential value of the Island's biodiversity and its ethnobiological richness. An interesting perspective on the challenge implicit in this work was provided by Tony Whitten (personal communication, 1992), who had spent longer on Siberut than any other ecologist. He observed that all of Siberut's obvious natural resources had already been removed by over-harvesting timber, rattan, and gaharu, and that attempting to develop the island was therefore 'like expecting a spider to run around after having six of its legs pulled off'.

Box 4.1 Aims of the Siberut project (from ADB, 1992b)

(a) Legally establish and physically demarcate the new nature reserve (the precise location, size, and shape of which was yet to be determined).

(b) Introduce infrastructure, personnel, and equipment to protect and manage the park in co-operation with the local people.

(c) Survey, map, and zone the island as a whole to define constraints and opportunities for socioeconomic development in harmony with the indigenous culture and living resources.

(d) Establish a permanent system for communication amongst the park and project managers and the local residents.

(e) Use that system to involve the local people in programmes for enhancing social welfare, including health care, literacy, environmental education, and village participation in rattan planting and marketing and in ecotourism development.

(f) Actively restore sites within or outside the park which have been seriously damaged by logging activities prior to their cessation.

(g) Establish a biodiversity research station in or near the park, as a contribution to the systematic documentation and management of Indonesia's national biodiversity assets.

(h) Undertake further studies and monitoring activities to establish what more, if anything, needs to be done to maximize Siberut's long-term and sustainable contribution to national development.

4.2.2 The Ruteng project

The design of the Ruteng project started from the assumption that its main focus would be to consolidate all natural forests in the area within a single nature reserve, either a Grand Forest Park (*Taman Hutan Raya*), or a Natural Recreation Park (*Taman Wisata Alam*). These categories of protection under Indonesia's Act No. 5 of 1990 both recognize that the area's main function is to preserve biodiversity, while allowing some use for research and tourism. Protecting the area from fire, encroachment, and logging would be a high priority in either case. Protecting the forests would mean also helping the people living closest to the new reserve to meet their development needs by improving agriculture and agroforestry, without cultivating inappropriate areas. The general demand for firewood and construction timber in the area could not continue to be satisfied by exploiting natural forests, and a community-based timber plantation programme was proposed instead. Since terracing, irrigation, and tree-planting were all already well established locally, little difficulty in achieving these objectives was anticipated.

The aims of the Ruteng project were therefore rather similar to those of the Siberut project, as adapted to local conditions. They included establishing and demarcating the nature reserve, and introducing infrastructure, personnel, and equipment to protect and manage it in co-operation with the local people. The whole area was to be surveyed, mapped, and zoned, and a village-based consultation system would be put in place around the new reserve. That system was to be used to involve the local people in programmes for enhancing social welfare and stabilizing land use, including health care, literacy, environmental education, and village participation in tree-planting programmes and ecotourism development. Sites within or outside the new park which had been seriously damaged were to be restored, a biodiversity research station was to be established, and further studies and monitoring were to be undertaken to identify unmet needs and opportunities.

4.2.3 Implementation arrangements

The Siberut and Ruteng projects were designed to achieve a balance, or at least a compromise, between large-scale land-use planning and small-scale, community-based development. The fact that large, centralized institutions were involved imposed one constraint on this, since they required top–down planning in accordance with existing procedures. The cross-sectoral aims of the projects imposed another, since ways would have to be found to achieve efficient consultation among various government agencies. At the national level, a committee chaired by the National Development Planning Agency (BAPPENAS) would provide strategic oversight and policy direction, while regional equivalents would exist at the Provincial and District levels. These were to be based closer to the project area, where they could provide more local oversight and routine coordination. The Ministry of Forestry would be involved at all levels through PHPA.

The main contract to provide all national and international consulting services to PHPA, and to manage with PHPA all aspects of the implementation of the project in Siberut, Ruteng, and at PHPA headquarters was awarded in late 1993 to an American consulting firm, Chemonics International. Shortly afterwards, the Chemonics project manager (R. Pardo, personal communication, November 1993) observed that the original project design appeared 'to follow current thinking regarding the importance of local participation, in which case the project may require PHPA to depart from its traditional way of doing business and learn to become partners with local communities and indigenous peoples in the management of the project areas (protected areas and buffer zones). I think our biggest challenge will be to develop a process that will facilitate the creation of such partnerships, and more importantly, sustain them after the project terminates.'

This approach reflected efforts during the pre-appraisal (ADB, 1992b) to find ways to implement the project in collaboration with local NGOs in each of the project areas. It was intended to rely on local NGO services as the primary means to deliver community awareness, mobilization, and extension programs (CAMEPs) in each target area. Each CAMEP would emphasize low-key but continuous encouragement of environmental education, awareness-raising, and participatory planning at the village level. The idea was to provide these inputs through an integrated programme reflecting these linked themes and based on an intimate, long-term relationship between local people and NGO and project personnel living among them. It was thought that the three elements of technical advice, physical resources, and financial resources would need to be combined to make a successful CAMEP, and to be presented in a way acceptable to target communities and individuals. The need for technical advice was to be addressed through the development of suitable extension and training packages by specialist staff working in the sites, while resources were to be provided in accordance with the project's budget.

The final element, presentation, was to involve using part-time project employees at the village level, to act as long-term contacts for the project. This was to be arranged by contracting a local NGO to recruit, train and manage local people at the level of *Rukun Warga* (*ca*. 50 households) and *Desa* (*ca*. 500 households) in each project area. The role of these 'development assistants' was to be to help local people understand the aims of the project and to help the project understand the needs of the people, while also helping each community to prepare its own local development plan. The aim was that the project would then be able to respond to proposals for assistance based on the development plan.

Training and communications in the CAMEP were to use workshops, meetings, and teaching or extension materials as needed by the communities and the project. These were all budgeted within the project when it was redesigned at the pre-appraisal stage, as were the various physical resources which were expected to be needed to effect change at the village level (e.g. health-care supplies, agroforestry seedlings). In the Siberut and Ruteng sites, it was intended that the CAMEP would be preceded by a one-year preliminary awareness programme, the aim of which was to sensitize local communities to aims of the project, while also identifying the first group of local recruits for employment under the CAMEP. This phase was to coincide with the process of surveying and mapping each project site, since these operations would involve contact between surveyors and resident communities with important public-relations implications. The project was reviewed by the ADB in March 1994, and found to be strongly supported by local government and especially by local communities and NGOs (ADB, 1994; Box 4.2).

Box 4.2 Summary of the ADB project review mission
(quoted from ADB, 1994:1).

'1. During the Review Mission in March 1994 it was noted that the support and enthusiasm about the Project expressed by the local government authorities at the district level, the local NGOs and the community leaders at the Project sites, has been impressive. The staff assigned to the Project has also shown good understanding and willingness to undertake the complicated task of implementation. The group of young men and women foresters assigned to the field Project sites are all quite energetic and willing to take up their assignments in remote locations. These persons, however, would need to be motivated and provided with appropriate incentives, given the difficult working and living conditions at the Project sites.'

'2. Of special significance was the level of awareness exhibited by the local indigenous communities at the Project sites, brought about through active participation of the local NGOs. The community leaders of Mentawai and Manggarai communities openly displayed their enthusiasm and commitment to the Project, especially the aspects that dealt with improving their lot and alleviating their poverty through development activities aimed at improved farm production, employment generation, rattan production and processing, handicraft development, improved marketing of the local products, increased availability of fuelwood (Ruteng), and others, planned and implemented through their active participation. They appeared overwhelmed by the Government's commitment to solicit their support and participation in planning and implementing various Project activities, including demarcation of boundaries. The local NGOs, conveying the aspirations of the local communities, pledged their total cooperation and support for the Project.'

'3. Physical process of implementation to date is admittedly slow (mainly due to initial delays in mobilization, limited familiarity of the PHPA with Bank operations, and administrative delays in engagement of consultants) but appears steady and proceeding satisfactorily. Significant achievement in terms of mobilization, engagement of consultants, survey and mapping, engagement of local NGOs, and the initiation of the Preliminary Awareness Program, goes to the credit of the Project Management, both at headquarters as well as at the Project sites.'

4.2.4 Supportive measures

The pre-appraisal of the project specified work to be undertaken at PHPA headquarters, including a variety of studies, training programmes, and institutional improvements to allow the lessons learned from the field projects to be applied within the Indonesian reserve system as a whole. At the national level, the project was to provide technical assistance to assist PHPA in guiding and overseeing activities at the project sites, and in providing advice on conservation policy, monitoring systems, institution building, and human resource development to support PHPA's long-term development.

These measures were intended to relieve some of the handicaps under which PHPA was working (Caldecott, 1992a). Firstly, PHPA did not then possess an accurate, updatable monitoring system for mapping and keeping track effectively of the reserves in its charge. Secondly, it did not possess nearly enough human or material resources to keep watch on all of those areas or to protect them from harm. Lastly, there was no reliable administrative mechanism to ensure communication amongst or to resolve conflicts of interest between PHPA and other groups wishing to use the reserve system destructively. Put more simply, PHPA was finding it hard to tell exactly where its main assets were or what was happening to them. It often could not intervene to protect them, and if it tried to do so it could be overruled by other interest groups. By 1994, the groundwork for resolving some of these problems was being laid under the auspices of the ADB biodiversity project (e.g. British Council, 1994).

4.3 Discussion

The *Biodiversity Conservation Project in Indonesia* sought to solve conservation problems in two very different locations. Siberut was a rain-forested island with a history of large-scale logging and inappropriate resettlement, while Ruteng was a semi-evergreen moist forest area, vulnerable to fire and with a history of small-scale logging and agricultural encroachment. Siberut was a remote, sparsely settled island, while Ruteng was a key watershed overlooking a town. To address these issues within one project was to confront in microcosm the main dilemma of PHPA itself. This was the question of how to find ways, within a centralized system of government, to apply locally appropriate solutions to site-specific problems throughout a large and very diverse country.

An important breakthrough was achieved in designing the project to rely strongly on the involvement of local NGOs, and in persuading PHPA and the Indonesian government to accept this design. Problems in implementing these arrangements are likely, but the key principle has now been established that managing Indonesia's reserve system will require formal but flexible local partnerships between government and people, rather than a continued policy of central management with inadequate governmental resources.

The administrative history of the ADB project is also interesting, since in the 1990s the National Development Planning Agency (BAPPENAS) had increasingly tended to pair donors with nature reserves, so that once an area was adopted by a donor, others were encouraged to look elsewhere for project sites. This had the virtue of dispersing technical and financial assistance across the reserve system. It would also make it possible to compare results between projects, where these were managed by different donors with different approaches and procedures. A review of all the projects in the future should thus allow important conclusions to be drawn, about which elements in project design and implementation are most effective in conserving biodiversity and assisting national development.

As originally conceived, the ADB project was to have comprised work in the Togian Archipelago in Central Sulawesi and the Gunung Lorentz area of Irian Jaya, as well as in Siberut and Ruteng. The cost of a project covering all four sites was anticipated to exceed US$100 million. During project processing, the ADB decided to adopt only two sites, Siberut and Ruteng, based mainly on the need for an urgent response to the threats to them. Another factor was that the ADB wished to reduce the scale of the project so as to be confident of having the capacity to implement it successfully.

This caution reflected the fact that this was the ADB's first major involvement in the complex field of biodiversity conservation, and it was intended to consider adopting other areas in the future once experience had been gained from work in Siberut and Ruteng (B. N. Lohani, personal communication, July 1994). With respect to the Togian Archipelago, there was also a conscious decision to defer ADB assistance until more resources to manage marine parks effectively were available. Provision for marine training was therefore included in the project design, and was also a response to PHPA's commitment to manage 300 000 km^2 of marine and coastal reserves by the end of the century.

5

Forest fragments in China and the Philippines

5.1 Introduction

An increasing number of tropical countries have only small areas of remaining natural forests, distributed in small patches (Collins, 1990; Collins, Sayer, & Whitmore, 1991; Sayer, Harcourt, & Collins, 1992). Conserving their native species has thus often come to mean working with fragments of habitat and small, remnant populations. This work must usually be done against the strong social and economic pressures which caused widespread deforestation in the first place, and which are often still active around each nature reserve (Laidlaw, 1994). Conservationists must also meet and somehow overcome the challenges posed by ecological and genetic constraints. Minimum populations must be maintained if wild species are not to drift into extinction, and minimum areas of habitat must be retained to support those populations (Diamond & May, 1976; Wilcox, 1980; Frankel & Soulé, 1981; Lovejoy et al., 1986; see Chapter 9).

Many tropical forest species occur at low population densities, so large areas of habitat are needed to support viable samples of their populations. A rough guideline is that a tropical forest reserve should be 500–1000 km² in area if most native species are to survive there in the long term (Myers, 1986). Hainan Island in South China, and southern Luzon in the Philippines, are two places where deforestation and forest fragmentation are far advanced, and which possess large numbers of species, many of which are endemic. Projects in these areas have no option but to try to reconcile human need for space and other resources with the constraints of ecology and population biology.

Endemic and endangered Philippine swallow-tail butterfly,
Graphium idaeoides, as mentioned in the text.

5.2 Hainan Island, China

5.2.1 Conservation issues

China's land area of 9.3 million km^2 encompasses a huge range of natural
ecosystems, from the grasslands, deserts, and coniferous forests of central
Asia, to the mountains of the Himalayan Plateau and Tibet, and the temperate,
sub-tropical and tropical lowlands of the east and south. Nearly 1.2 billion
people were living in China in 1993, or 21% of the world's total population,
and this number was increasing at a rate of about 1.2% each year (Dompka,
1994). These great numbers of people have placed many demands on China's
environment, and much of the country has been deforested in meeting the
needs of subsistence, revolution, war, agriculture, and industrial policy (Smil,
1983, 1984, 1992; Li *et al.*, 1988; Richardson, 1990; World Bank, 1991b;
Schaller, 1993).

In 1979, China began a process of reform aimed at promoting rapid economic growth, which resulted in an average growth in national GDP per person
of almost 8% per year in 1980–90 (UNDP, 1993). The twin policy aims of
deregulation and decentralization were advanced by creating a number of
Special Economic Zones (SEZs). The largest of these was set up in 1988 to
cover the whole island Province of Hainan, which lies about 30 km off the

southern coast of Guangdong Province, in the South China Sea (Fig. 5.1). Since becoming an SEZ, Hainan has experienced rapid industrialization and urbanization, and the island's GDP grew by more than 10% annually in the early 1990s (Holdgate, 1993).

Hainan has a land area of almost 34 000 km², about half of which comprises a flat coastal plain rising to rolling low hills. There is a mountainous interior, with 667 peaks over 1000 m, the tallest of which is Wuzhishan at 1867 m. The annual rainfall is about 2500 mm in the north-east, but due to prevailing winds and rain shadows only about 1000 mm in the extreme west. Most rainfall occurs from May to October, when frequent typhoons also arrive from the Pacific. Hainan is at about the same latitude as Hawai'i and its climate is tropical, although temperatures sometimes briefly approach freezing. Varied growing conditions on the island result in a mosaic of distinct natural ecosystems from mangrove and beach forests, through various types of lowland forest to the tall and short forms of montane forests and elfin woodland on exposed mountain crests above about 1600 m (ADB, 1992c).

Hainan is very species-rich, and its isolation from the mainland has ensured high levels of endemism. There are up to 4500 higher plant species of which about 600 are endemic, and the island is listed as a centre of plant diversity (WWF & IUCN, 1995). Vertebrates include 37 amphibians and 104 reptiles, including the endemic Hainan tortoise (*Chuora hainanensis*), and at least 344 birds of which enough have restricted ranges for the island to be listed as an endemic bird area (ICBP, 1992, 1995). Three of the 76 mammal species are endemic, these being a hare (*Lepus hainanus*), a moonrat (*Neohylomys hainanensis*) and a flying squirrel (*Petinomys electilis*). There are also 29 endemic subspecies, including the Hainan gibbon (*Hylobates concolor hainanus*) and the slope or brow-antlered deer (*Cervus eldi*). The latter had been reported extinct in China (Collins, Sayer, & Whitmore, 1991), but it has been bred in Hainan at the Datian National Nature Reserve, where in 1991 it was being used to make deer-bone tonic wine (ADB, 1992c). Little is known of the fresh-water fish and invertebrate faunas.

About 6.6 million people were living in Hainan in 1990, mostly of the Han (83%), Li (15.6%), and Miao people (0.8%), but also including a few thousand Zhuang and Hui (ADB, 1992c). The Li and Miao arrived 3000–4000 years ago, and were followed by successive waves of Han immigrants who colonized, cleared, and cultivated the lowlands. The Li and Miao were thus displaced into the mountains of the south and west, where they continued to live by shifting cultivation. This pattern is similar to that elsewhere in south and interior China, with Han Chinese dominating the lowlands, and

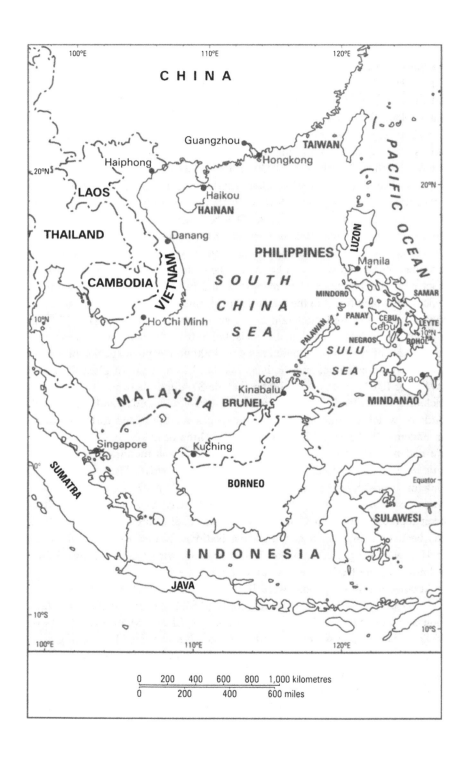

minority shifting cultivators occupying the mountains. Since most of China's lowlands have been settled and farmed for centuries, it is in the highlands where most natural forest and wild species remain.

When the People's Republic of China was established in 1949, about 8600 km², or 26% of the land area of Hainan, was under natural forest cover. Since then, the normal pattern in Hainan has been for natural forests to be logged, cleared, and replaced by plantations of tropical tree crops such as rubber (*Hevea brasiliensis*), or by trees grown for wood fibre (such as species of *Eucalyptus*, *Homalium*, *Casuarina*, *Pinus*, *Tectona*, and *Terminalia*). The alternative fate was for natural forests to be degraded by shifting cultivation, which was being practiced at increasing intensity as local populations grew. These two processes had resulted, by mid-1991, in natural forests occupying only about 11.4% of Hainan's land area, while planted forests occupied about 26.5% (ADB, 1992c). Moreover, most of the surviving natural forest had been degraded, and intact forests occurred in small fragments amounting in total to only about 4% of the land area (or 1360 km²). The rest of the natural forest estate was a patchwork of regenerating forest and scrub, while all other lands on the island were under grass (often *Imperata cylindrica*), crops, or settlements, or else were entirely barren and in many cases actively eroding.

By the early 1990s, the scale of environmental degradation meant that a large and increasing number of Hainan's native species were under threat of extinction. Almost all the remaining natural forest patches were located in the mountainous southern half of the island, so lowland species and those confined to habitats occurring only in the north and east were particularly endangered or already extinct. A system of nature reserves had been established, of which 33 were in the interior of the island, with a total area of about 1273 km², or 3.8% of land area. They comprised 19 reserves (864 km²) dominated by intact forest, 8 (388 km²) dominated by disturbed forest, and 6 (21 km²) containing other natural ecosystems (Fig. 5.2).

The largest of the intact reserves was Yinggeling at 212 km², and only 2 others were larger than 100 km² (Wuzhishan and Jianfengling), while another 9 were between 20 and 60 km² in area. The larger reserves were all mountainous, and so contained only a limited sample of Hainan's flora and fauna, while the other ones were far too small to achieve the permanent protection of all the species and ecosystems within them. Most species had greatly reduced

Fig. 5.1 The South China Sea: Hainan and the Philippines

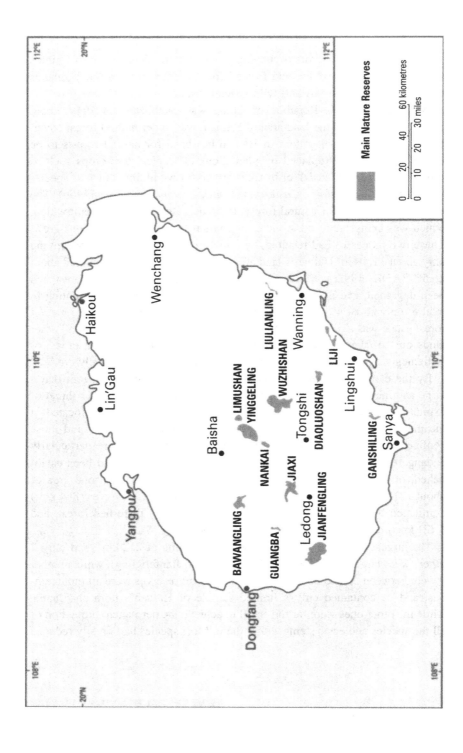

Main Nature Reserves

| 0 | 20 | 40 | 60 kilometres |
| 0 | 10 | 20 | 30 miles |

population sizes, and those small populations were becoming increasingly isolated as natural habitats continued to be lost outside the reserves, and as people exploited or encroached upon the forest. The biodiversity of Hainan was caught between the need of rural people for space and resources to live in traditional ways, and the demands of government for space for plantations and infrastructure.

5.2.2 Designing responses in Hainan

Before 1991, there had been little attention to the strategic conflict between officially planned and traditional land uses in Hainan. Nor had much thought been given to the impact on the environment of rapid industrial and urban expansion and the promotion of plantation agriculture and mass tourism. These events were occurring without any means to anticipate and manage environmental impacts, or to resolve conflicts between groups making incompatible demands on resources. As the Hainan Provincial Government began seeking investment for the Special Economic Zone from external agencies, however, the lack of environmental controls began to raise questions over the sustainability of the Province's development strategy. Recognizing this, groups within the government set up an International Advisory Council (IAC) to provide guidance on environmental matters, which it did with the assistance of the World Conservation Union (IUCN).

The IAC helped to process a government request for assistance from the Asian Development Bank, which commissioned an environment and natural resources masterplan for Hainan (ADB, 1992c). This work was undertaken by Environmental Resources Management Ltd, and its aim was to study the capacity of Hainan to achieve sustainable development. The masterplan was linked to an action plan, which aimed to define specific projects to avoid and manage environmental problems, and to use natural resources sustainably to enhance the well-being of Hainan's population (ADB, 1992d). This was to be based on the use of physical, social and economic resources, including the Province's waters, lands, forests, coastal environments, and components of its biodiversity.

Fig. 5.2 Nature Reserves of Hainan

These projects included several which aimed to apply the concept of IUCN's Category V Protected Area, and therefore to create 'protected landscapes' in Hainan. This category of reserve is defined to include regions where the dominant management aim is to maintain environmentally sound or unique patterns of human settlement and use of living resources (Lucas, 1990). Such areas are usually envisioned to incorporate agriculture and local settlements, as well as sustainable forest and wildlife management projects, with recreation and tourism as important additional aims. They may include zones in other, more rigorous protection categories, provided environmental usage is integrated overall. A protected landscape may thus include areas under a variety of kinds of ownership, and because of this carefully integrated planning is required for each area as a whole.

This is the most flexible of IUCN's protected area categories, and seemed best adapted to the needs of several nature reserves, the species and ecosystems within them, and the people living around them. This applied particularly to areas where it was considered possible to secure species-rich low-altitude forests and the continuum of natural forests rising into the mountains above them. There were few opportunities to extend nature reserves from peaks and ridges downhill into contiguous but unclaimed forest land, and it was thought that a way round the problem would be to create special management zones around existing nature reserves. These zones would use the Category V concept, and seek to improve conditions for wild species in order to increase the size of the habitats available to them.

An essential component of this strategy was to meet human needs within those natural systems, and this would involve local consultation and participation in planning. Thus, the protected landscapes would function as biodiversity-oriented planning and management units, with control based on a formal partnership among all local stakeholders in the project area. The role of government would be to oversee the operation of the local partnership, to guarantee the permanence of biodiversity refuges within the protected landscape, and to arbitrate between conflicting claims on resources. This was a new direction for government in China, and would involve compromise and flexibility by government and local interest groups. Two protected landscape projects were designed, at Jianfengling and Wangxia.

(a) The Jianfengling project

The project area comprised a cluster of granite peaks up to 1412 m high in the south-west part of Hainan (Fig. 5.3). The topography and soils were complex, and natural ecosystems were diverse and species-rich (Liu, 1992; Holdgate, 1993). The area was one of Hainan's most important remaining forest regions, but its resources were under demand by a several different users, including:

- the Forest Bureau, which managed the 107 km² Jianfengling nature reserve;
- a Provincial Subordinate Forest Company, which had a 472 km² concession, of which about 65% was natural forest;
- a State Farm;
- the Chinese Academy of Sciences, which had an experimental station nearby; and
- the military, which had two reservations inside the buffer zone of the nature reserve.

Other forest lands existed outside the concessions and had been exploited or modified to various degrees, and there was agricultural encroachment on the fringes of the Jianfengling area which graded downhill into settled regions in the surrounding lowlands. All these categories of forest land together amounted to an area of about 1000 km², all of which was to be affected by the Jianfengling project. This project was designed to take account of the area's outstanding biodiversity and its importance in watershed protection, as well as its potential as a highland tourist resort, and the fact that it was also the site of a major forestry institution and a number of agricultural concerns. All of these were competing for resources in the same area, most of which was steep and fragile, and this seemed unlikely to have a sustainable outcome. The aim of the project was therefore to simplify and focus the objectives of land and forest use so as to avoid competition and improve management within the Jianfengling protected landscape.

This would be placed under the control of a single, local management committee comprising representatives of all groups with an interest in the future of the area. The whole project area would then be surveyed and mapped, and land capabilities assessed in detail. Using this information, and guided by the need to maintain watersheds and to preserve biodiversity, the whole project area was to be divided into zones where different and complementary management priorities would be proposed for discussion by the members of the management committee. A draft management plan would be prepared, making specific recommendations for the allocation of activities to particular areas, and providing detailed guidelines for those activities. The draft plan would be reviewed and amended by the management committee before being approved as the guiding document for all development in the project area.

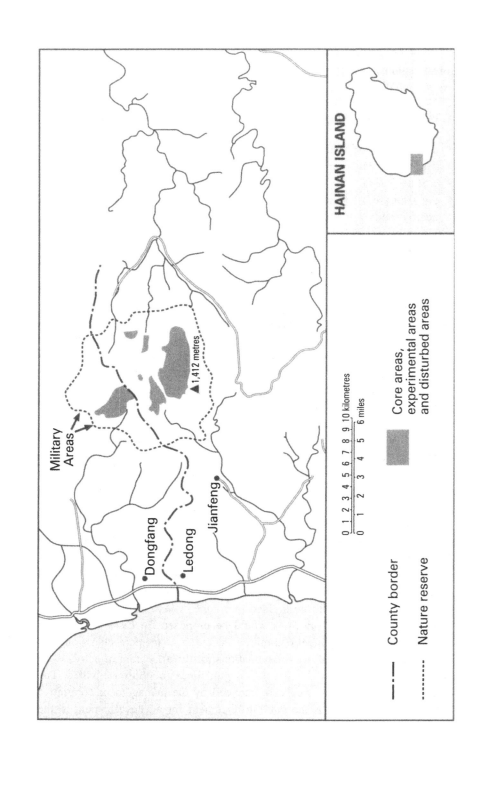

Military
Areas

Dongfang

Ledong

Jianfeng

1,412 metres

0 1 2 3 4 5 6 7 8 9 10 kilometres
0   1   2   3   4   5   6 miles

County border

Nature reserve

Core areas,
experimental areas
and disturbed areas

HAINAN ISLAND

Meanwhile, it was also necessary to help the forestry company to devise an environmentally sound business plan with which to attract investment in line with the management plan for the protected landscape. Investment would be needed to demarcate all zones of the protected landscape physically, and to monitor and police the use of land within those zones. The management committee itself would need to have sufficient power to arbitrate bindingly on all parties concerning the use of resources within the Jianfengling protected landscape. It would exercise this authority in the light of information provided by continuing studies, and in parallel with consultation among the members of the management authority.

(b) The Wangxia project
The project area was located inland to the north-east of Jianfengling. It comprised the Bawangling nature reserve to the north and the Jiaxi nature reserve to the south, and between them a nearly circular valley bounded by steep slopes and watersheds, through which intact forests extended between Bawangling and Jiaxi (Fig. 5.4). The floor of the valley was occupied by the small Li village of Wangxia and several settlements nearby. Biological links between the two nature reserves would depend on maintaining natural forest cover within the valley, and would therefore depend on land use by the people living within it. An aim of the Wangxia project was therefore to protect these intervening forests, and to try to make it possible for the local people to participate in the benefits of doing so. A thorough survey was proposed, followed by zoning, marking boundaries, and investing in managing, monitoring, researching, and developing the area under the authority of a local management committee.

The Bawangling reserve itself was uninhabited, but local people outside the Wangxia valley used it for hunting, rattan cutting and gathering medicinal plants. Encroachment and boundary changes during the 1980s had reduced its area to about 56 km², between 350 m and 1440 m in elevation. About 22 km² of this was intact montane forest, and the rest was a patchwork of logged oak and dipterocarp forest, forest enriched with planted lychee (*Litchi chinensis*), and regenerating forest, scrub, and grassland partly planted with

Fig. 5.3 The Jianfengling project area

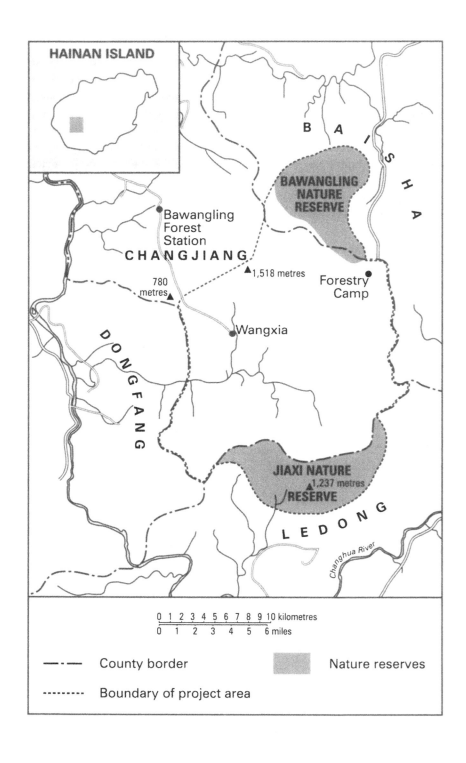

HAINAN ISLAND

B A I S H A

BAWANGLING
NATURE
RESERVE

Bawangling
Forest
Station

CHANGJIANG

▲1,518 metres

Forestry
Camp

780
metres▲

Wangxia

D O N G F A N G

JIAXI NATURE
▲1,237 metres
RESERVE

L E D O N G

Changhua River

0 1 2 3 4 5 6 7 8 9 10 kilometres
0   1   2   3   4   5   6 miles

—·— County border

········· Boundary of project area

▮ Nature reserves

fir and pines. The Bawangling Provincial Subordinate Forestry Company had a concession near the reserve about 637 km^2 in extent, of which about 212 km^2 was natural forest. The forestry company thus represented a major influence on the future of land use in the project area, and its investment strategy and approach to forest management were thought to be relevant to the project.

The Jiaxi reserve, on the far side of the Wangxia valley, was about 60 km^2 in area, and although it was under some pressure by hunters and small-scale loggers, its forests mostly remained intact (ADB, 1992d). The area contained an important montane forest system which would be particularly significant in view of its connections with the Wangxia valley and with Bawangling to the north. This made it desirable to integrate the management of Jiaxi with that of the larger area as a whole. The key to this was maintaining natural forest cover within the Wangxia valley, and this meant responding to the needs of the local people there in ways which would safeguard the forest. Two options were proposed, involving their long-term participation either in the protected landscape project, or in a resettlement programme.

Resettlement was considered for two reasons, the first of which was that the valley was isolated, and the local people had already requested improved road access and development assistance from government. If this was provided there was a strong likelihood that pressures on land and forest within the valley would increase sharply. Secondly, the condition of land in nearby areas outside the valley suggested that traditional shifting cultivation had exceeded the ability of the land to support it, since steep slopes had been cleared and were dominated by *Imperata* grassland, and in places were scarred by gully erosion and landslips.

The question seemed to be whether or not it was possible to satisfy the expectations of the people of the Wangxia valley within the valley itself, without losing the opportunity to conserve an important part of Hainan's biodiversity. This would mean their abandoning shifting cultivation on steep land and intensifying the use of valley bottoms, and would need to be supported by comprehensive education and forest protection programmes. This was considered to be feasible but expensive, and it contained an element of

Fig. 5.4 The Wangxia project area

risk because it would entail improved access and population pressures in the middle of the nature reserve complex. The alternative of resettlement was preferred by the State Science and Technology Commission, in view of the uncertainty of other measures working in the long term, and the large numbers of endangered and endemic species within the Bawangling and Jiaxi reserves (Holdgate, 1993).

It was also necessary to find ways to assist the development of the Bawangling forestry company and to stabilize land use around the outer fringes of the reserve system. For these purposes the project area was to include the forestry company's concession and all other areas where land and forest use were likely to affect the Wangxia valley or the two nature reserves, making an area of about 800 km^2 in total. The methods proposed for the Wangxia project were thus similar to those for the Jianfengling project. They were to involve the creation and empowerment of a local committee with inclusive local membership, followed by a detailed survey and mapping exercise, zoning of the area, and development of a management plan based on consultation with all actual or potential users. Meanwhile, the local forestry company was to be helped to develop an environmentally sound business plan, with which to attract investment in line with the management aims of the protected landscape as a whole.

5.3 Makiling Forest, the Philippines

5.3.1 Conservation issues

The Philippines is an archipelago of about 7100 islands, with a total land area of about 300 000 km^2, and a coastline of more than 34 500 km or 6% of the world's total. It is located between Borneo and Taiwan, and between the South China Sea and the Pacific Ocean (Fig. 5.1). The two largest islands are Luzon and Mindanao, which account for 68% of total land area. Most of the islands have mountainous interiors, with lowlands limited to narrow coastal plains. The climate is strongly seasonal, and the timing and amount of rainfall depends on location within the archipelago, and on rain-shadow effects from the mountains relative to rain-bearing monsoon winds. Much of the country is also affected by typhoons, which usually arrive from the Pacific between July and November.

The Philippines had a population in 1993 of about 65 million, which was increasing at 2.5% annually, and with about 217 people km^{-2} it was one of the most densely settled countries in South-east Asia (Dompka, 1994). It was colonized by the Spanish, early in the sixteenth century, who introduced

Roman Catholicism and the practice of *hacienda* farming and ranching as they did in Central America. This resulted in much of the country being occupied by large estates owned by élite families of Spanish descent, and in chronic land hunger among the autochthonous population. This landlessness contributed eventually to the growth of large cities, such as Metro-Manila in Luzon, and to mounting pressure on publicly owned or unoccupied forest land for farming. In the absence of effective land reform before or during the Marcos dictatorship up to 1986, a Communist insurgency became established in much of the countryside, led by the New People's Army (NPA).

Comprehensive land reform was attempted during the Aquino Presidency of 1986 to 1992, but there was strong resistance by influential land-owning interests, and the tendency was for only public forest lands to be redistributed to the rural poor, thereby accelerating deforestation. An important result of this history of land hunger and limited land reform, and the ideological conflict resulting from it, was that the issue of access to land became highly politicized. Many rural people and intellectuals became convinced that poor people have the *right* to farm in forest areas, regardless of the environmental consequences, and this view has been broadly supported by government policy since 1986. With the NPA waiting to exploit dissatisfaction, it was hard in this period to take a firm line on forest protection, since this could be seen as a return to the 'Marcos way of doing things'.

Part of the 'Marcos way', however, was logging, initially for export at a rate of about 10 million m^3 yr^{-1} until 1974, and then, following export restrictions increasingly for local processing and use (Boado, 1988). Illegal log exports remained common, however, and involved about one million m^3 yr^{-1} in the mid-1980s (Callister, 1992). Total recorded hardwood production declined steadily during the 1980s, from over 6 million m^3 in 1980 to less than 2 million in 1991 (FAO, 1993). This was partly in response to government logging bans, which by 1989 applied to all 64 provinces with less than 40% tree cover (out of 73 in total), all virgin forests on steep land, and all montane forests. In 1992, logging bans were extended to all virgin and high-altitude forests and all forests remaining on steep lands. Despite these efforts, logging and farming reduced forest cover in the Philippines from about 105 000 km^2 in 1969 to about 65 000 km^2 in 1988, or about 22% of total land area, and deforestation has since continued at a high rate (Cox, 1988; Collins, Sayer, & Whitmore, 1991).

The surviving flora of the country is highly diverse, with at least 8000 species of flowering plants and 900 ferns, and an estimated 10–15% more as yet undescribed (WCMC, 1992). About 3500 of these species are endemic (Madulid 1982; WCMC, 1994). About 960 terrestrial vertebrate species are

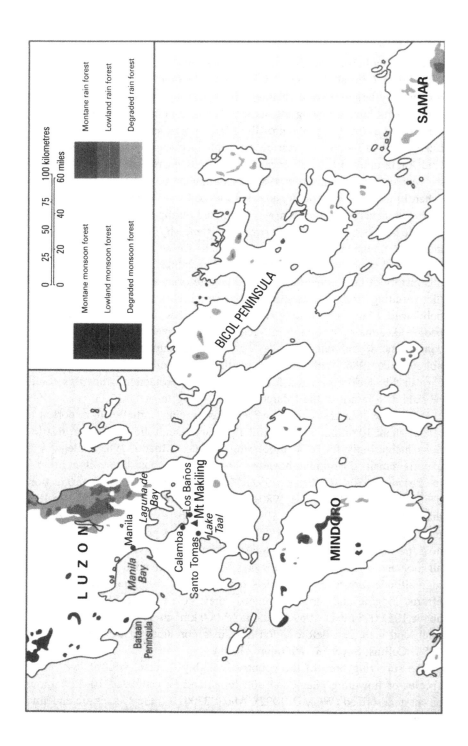

	Montane monsoon forest		Montane rain forest
	Lowland monsoon forest		Lowland rain forest
	Degraded monsoon forest		Degraded rain forest

0 25 50 75 100 kilometres
0 20 40 60 miles

LUZON

Bataan
Peninsula

Manila
Bay

Manila

Laguna de
Bay

Calamba
Los Baños
Santo Tomas Mt Makiling
Lake
Taal

BICOL PENINSULA

MINDORO

SAMAR

known to occur in the Philippines, of which up to 50% occur nowhere else (Heaney 1986; Cox 1988; Petocz, 1988). There are 160 or more species of mammals, 100 of which are endemic to the country and 45 are endemic to single islands within it. There are at least 556 bird species, of which 395 are breeding residents and 183 of these are endemic. Among reptiles and amphibians, there are 253 species, with endemism rates of 75% for lizards, 52% for snakes, and as much as 70% for amphibians.

As in Hainan, the biological richness of the Philippines, combined with widespread destruction of natural ecosystems there, means that all surviving areas of natural habitats and populations of native species have an exceptional conservation value. Southern Luzon is particularly badly deforested, and one of the few remaining patches of native vegetation there is on Mt Makiling (Fig. 5.5), a dormant volcano about 65 km south-east of Manila. It is mostly steep and rugged, with six drainage systems radiating from the summit at 1050 m, and flowing into densely settled and cultivated lands in the plains below. The climate of southern Luzon is moist but seasonal, with a mean annual rainfall of about 2400 mm, most of which falls during the south-west monsoon in May to December. Even during the drier north-east monsoon, however, the upper levels of Mt Makiling remain exposed to cloud, and above about 750 m the forest is mossy and continuously moist.

The forested peak and slopes (down to about 400 m in the west and south, and below 200 m in the east and north) lie within the 44 km^2 Makiling Forest Reserve. The Reserve thus includes a complete sequence of altitudinal forest types from lowland to upper montane ecosystems, including mixed diptero-carp forest (below about 500 m), with a canopy 25–40 m high and occasional taller emergents; lower montane rain forest (500–900 m), with a canopy 20–25 m high and dense understorey vegetation; and upper montane rain forest (above about 900 m), with stunted and gnarled trees up to about 12 m in height and with abundant epiphytic and other mosses. The flora of the mountain comprises at least 2238 species of vascular plants, including 638 tree species. There are 13 species of mosses and 29 ferns which are endemic to the Mt Makiling area, while 60% of all known Philippine fungi occur there (Pancho, 1983). Although much taxonomic work is still required, it is clear that the Makiling flora is very diverse, and that a large proportion of it is unique to the area. The lowlands and small, isolated mountains of southern

Fig. 5.5 Southern Luzon and Mt Makiling

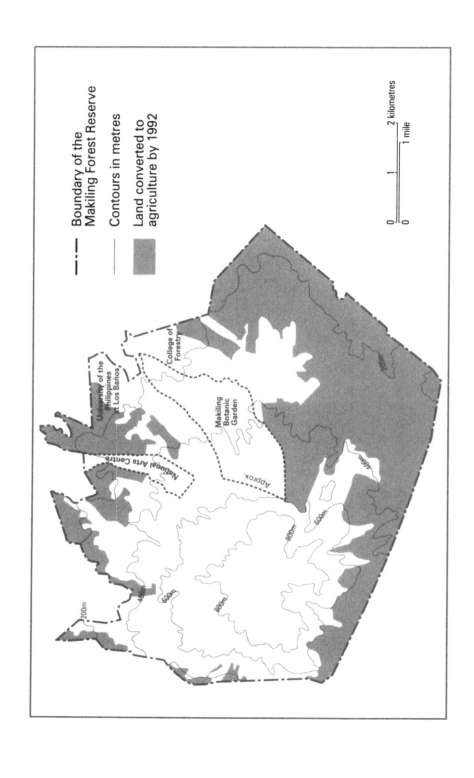

Boundary of the
Makiling Forest Reserve

Contours in metres

Land converted to
agriculture by 1992

University of the
Philippines
at Los Baños

College of
Forestry

National Arts Centre

Makiling
Botanic
Garden

Approx.

200m

400m

500m

800m

900m

600m

400m

200m

0   1   2 kilometres
0   1 mile

Luzon are listed as an endemic bird area by ICBP (1992), to which 13 restricted-range bird species are confined.

The vertebrate fauna of Mt Makiling includes at least 50 species of mammals and 120 birds, but the other groups are poorly known. The invertebrates of the mountain are thought to include at least 30 000 species of insect, but a majority of these and other arthropods have yet to be described (W. S. Gruezo, personal communication, February 1993). The extent to which this fauna is endemic and endangered is likely to be significant. An example is provided by the swallow-tail butterfly *Graphium idaeoides*, which is limited to low-elevation (below 330 m) rain forests on several Philippine islands where virtually no such habitat remains, including southern Luzon (Collins & Morris, 1985). If it is still present in the lower forests of Mt Makiling, it would represent the many invertebrates for which the Makiling Forest Reserve is perhaps the final refuge.

This Reserve was established in 1910, and was first administered by the Bureau of Forestry, as a forest reserve until 1920, then as a national botanic garden until 1933, and finally as a national park until 1952. The managing agency changed in 1952 and again in 1960, but it continued as a national park until 1963. Thereafter it was held by the University of the Philippines until 1987, when it was transferred by the Aquino government to the National Power Corporation for use as a geothermal energy source. It was returned to the university system in 1990, and placed under the control of the College of Forestry of the University of the Philippines at Los Baños (UPLB), with a mandate to use it for scientific and educational purposes (Caldecott, 1993a).

Throughout this varied administrative history, except for 1987–90, the main management aim for the Reserve was the protection of its biodiversity and its physical environment, and no large-scale exploitation or settlement was permitted. The repeated changes in management authority, however, had an effect on the continuity of protection programmes, as well as on the way in which the Reserve was perceived by local people. The degree of official tolerance for settlers in the Reserve varied over the years, and by 1993 the net result was that a gradual and politically irreversible influx of farmers had occurred. This had resulted in the conversion of 45% of the Reserve to farm-

Fig. 5.6 The Makiling Forest Reserve

land, and only about 20 km² remained as intact natural forest, almost a quarter of which was within the Makiling Botanic Garden (Fig. 5.6).

Most of these settlers originated locally and represented an opportunistic overflow of people from nearby lowland areas, taking advantage of access to free land, and often helped by roads built by the National Power Corporation to support geothermal drilling. By the end of 1992, almost 1000 people had established farms within the Reserve, of which about 20% lived within the Reserve itself and the rest lived outside it. By far the most common farming activity in the Reserve was agroforestry, involving diverse mixtures of fruit trees and non-woody crops in farms with an average size of between one and two hectares (Cruz, Francisco, & Torres, 1991; Dizon, 1992; Francisco, Mallion, & Sumalde, 1992; Mallion, Francisco, & Gagalac, 1992). This form of land use seemed to be well adapted to local conditions, and soil erosion rates on the farm holdings were similar to those expected under intact forest cover (P. Guy, personal communication, March 1994). This dominant land use resulted in an acceptable quality of life for most farmers, few complained of significant problems, and almost all of them intended to remain in the Reserve and pass their land on to their children. This relaxed attitude was a result of their experience of management agencies which, especially during the early 1990s, tended to avoid confronting those who had already settled, and instead focussed on trying to prevent the clearance of new farms in some areas, and on preventing the unlicensed cutting of wood.

Other factors affecting management of the Reserve in 1993 included the division of responsibility among several offices of UPLB. These variously maintained within its boundaries a botanic garden, long-term research concessions, and several areas which had been leased out by UPLB to other groups. These leaseholds confused tenure, since there were no definitive maps or agreements as to terms of occupancy. In addition, drillings by the geothermal plant were affecting part of the Reserve, but its operation was independent of the University and could not be supervised by UPLB. The result of inconsistent administration and protection over the years was that the Reserve had an unusually complex set of conservation problems. The most serious was that almost half of the Reserve was being farmed, in some cases by people who could trace their families' residence or use of land within the Reserve back 50 years or more. Another problem was that all of the natural forest which had not been converted to farmland was used on an open-access basis by harvesters of forest products, some of whom were residents and/or farmers, but many of whom were neither. Meanwhile, there were strong economic incentives for individuals to clear more land for agriculture within the Reserve, and this was occurring rapidly with pressures on both surviving lowland forests and on steep and highland areas. These pressures varied considerably among the six drainage systems of the mountain, due to differences in terrain, settlement patterns, and the accessibility of land still under forest.

5.3.2 Designing responses on Mt Makiling

(a) The biodiversity refuge
Groups wanting to use the Makiling Forest Reserve for purposes other than conserving biodiversity included various institutions and leaseholders, and both resident and non-resident farmers. A joint Canadian–Philippines project, administered by Dalhousie University, supported studies in 1993 which suggested that constructive action would be possible, but that it would need to be based on a much clearer vision of why and how to protect components of biodiversity in the Reserve (Caldecott, 1993a). It would also be necessary to find some means of paying compensation, in cash or kind, in order to make the measures required locally acceptable. By designing such a project, a model was being sought which might be more broadly applicable in the Philippines, where complex settlement pressures on remaining natural environments were common. The University would then be in a good position to ensure the spread of those techniques through its educational influence.

The original species richness of the Reserve seemed likely to have included several tens of thousands of species of plant and animal, mostly insects and other arthropods. The relative species-richness of lowland mixed dipterocarp forest meant that many native species would have been dependent on this forest type. The destruction of almost all lowland forests in southern Luzon meant that surviving forests around the base of Mt Makiling were already isolated. With little more than 4 km² of this habitat surviving in the Reserve, population sizes of most species were assumed to have been greatly reduced, and their genetic variation consequently lessened. Many were thought likely to drift into extinction in the wild, even if remaining intact forest was protected in the Reserve. They might therefore be numbered among the 'living dead', in which investment may well be pointless (Janzen, 1986a, 1988a).

If the wild species and natural habitats of Mt Makiling were to survive, a biodiversity refuge would be needed. This should contain as much lowland forest as possible, as well as contiguous forests uphill so as to include species adapted to conditions at different elevations. The only intact sequence of forest types on Mt Makiling ran from the edge of the UPLB campus, through the Botanic Garden and up to the summit, and the largest area (about 3 km²) of intact lowland forest was in and around the Botanic Garden. It was recommended that the biodiversity refuge should include this transect, securing about 8 km^2 of intact forest between about 100 and 1000 m elevation (Fig. 5.6). This refuge would be the main resource for further development based on the area's biodiversity. Other actions would then be needed to buffer and secure the transect, and to prevent further damage to the northern, western and southern slopes of Mt Makiling.

Creating the biodiversity refuge would involve stabilizing the boundary between land under natural forest and farmland. This boundary would need to be agreed by all parties and then clearly marked, regularly patrolled, and maintained in perpetuity. This basic distinction in land-use would need to be supported by accurate maps, based on those of land-holdings by Abraham, Sargento, & Torres (1992), which had established both the current boundary between forest and non-forest habitat, and also the basis for allocating tenure to individual farmers. Allocating such tenure would imply the acceptance that most areas which had been settled by farmers were not retrievable for conservation purposes, but would have allowed a permanent solution to be achieved, supported by a social consensus in favour of conservation.

(b) The role of the Botanic Garden

The Makiling Botanic Garden was regarded by UPLB both as a tourism asset and as a site for biodiversity research and teaching. Its development for tourism to some extent conflicted with its role in conserving biodiversity, since tourism projects underway in 1993 involved high-intensity use of the small area of lowland forest within the Garden. There was thus a need for a clear mission statement for the Garden, to clarify its role in conservation, education and tourism. It was recommended that the primary aim should be to use the Garden to save local species which would otherwise become extinct. Other roles and activities would then be adapted to complement that mission, including public education, applied research and collaboration with other botanic gardens and conservation groups. This would include developing tourism as a source of revenues in support of conservation programs in the Reserve.

A comprehensive inventory of biodiversity within the Reserve was proposed, partly to identify species at most risk of extinction. These were likely to include all those with very small population sizes in the Reserve, which could amount to several thousand plant and arthropod species. Since so much of the Reserve's total species richness resided in its invertebrate fauna, a shift in emphasis was suggested, in which the Garden would give greater attention to the fauna of the Reserve. This would move it in the direction of becoming a 'biodiversity centre' rather than a botanical garden in the traditional sense. This change in role would clarify and strengthen the link between the Garden and the biodiversity refuge of which it was to be such an important part.

Suggestions had been made in the early 1990s to develop the Garden into a national facility for off-site biodiversity conservation. The Garden had few resources for the task of saving local plants and animals, however, so such a change in role would have posed the danger of it being overwhelmed by accessions from all over the Philippines. It was therefore recommended that the Garden should not be given such responsibilities without also being given adequate resources with which to fulfil them, and that there should be no conflict with the protection of intact ecosystems on Mt Makiling itself. In any case, rather than creating a centralized national botanical garden, it was thought that a better solution would be a network of local off-site or near-site biodiversity conservation centres around the country, each specializing in its own local biota.

(c) Mobilizing public support

Obtaining public support for conservation on Mt Makiling was thought to require finding a way for the local public to participate in the benefits provided by the Reserve. A priority would be to establish permanent lines of communication between all parties with an interest in the Reserve. This would include employing and training local residents in farming communities in and around the Reserve. Thereafter, those communities would be helped to prepare their own plans for their own development. The aim would be for such plans to be in a form which could be used both as a basis for community self-help, and to guide the provision of assistance where possible and necessary.

A difficulty was that the University was perceived as an élite institution whose interests were remote from those of farmers in the Reserve. To change this, it was suggested that a broad-based body with management responsibility for the local environment would be needed, in which local residents would be able to feel fully involved and which they would therefore trust to resolve conflicts of interest over the use of resources. This body would need to include local NGOs, and also UPLB and other institutional stakeholders in the Reserve. One option would be to establish a non-profit foundation which would manage the Reserve and be jointly controlled by farmers' groups, other stakeholders, and the University itself.

In establishing such a foundation, special attention would need to be paid to the long-term resident settlers within the Reserve, who represented about 20% of the people who farmed there. The non-resident farmers and other users of the forest were a source of pressure on the Reserve but had little reason to manage natural resources sustainably. Their presence was therefore considered to be incompatible with protection of the remaining parts of the Reserve. It was therefore proposed that established farms in the Reserve should be surveyed, with most becoming the subject of conditional leasehold agreements. Residents would be allowed to remain, subject to their agreeing to restrictions on using pesticides on their farms and on their use of the remaining natural forests. As residents with established leasehold tenure, the farmers would be encouraged to participate in the management of the Reserve. They would also share in all benefits obtained through sustainable management of the Reserve by the foundation described above. These benefits would come initially from increased charges for access and use, including both recreational visitors and renegotiated leases affecting parts of the Reserve.

5.4 Discussion

Hainan and southern Luzon both remained extremely rich in species in the early 1990s, and many of these were endemic. Both were also seriously deforested, however, with a very large proportion of the native flora and fauna dependent upon the survival of patches of natural forest only some tens of km^2 in extent. The projects designed in response to these circumstances were intended to safeguard such patches in the context of intense pressures to use them unsustainably through logging, or to convert them to farmland. In Hainan, the aim of both the Jianfengling and the Wangxia projects was to create a larger effective patch size, by consolidating existing nature reserves with surrounding areas of natural forest, and then managing the whole area in ways sympathetic to the needs of native species and ecosystems. In the Makiling Reserve, the aim of the project was to create a biodiversity refuge and to give a reason to maintain its boundaries to all groups with a long-term interest in the Reserve.

The underlying problems which the projects were intended to address were similar in some ways and different in others. In the highlands of both Hainan and southern Luzon, economic and political processes had resulted in a conflict between land-hungry farmers and biodiversity managers, the latter being the provincial government in Hainan and UPLB in on Mt Makiling. There was thus a similar conflict of perception about what specific lands should be used for. Both sides were perhaps more sophisticated in the Philippines than in China, since the conflict was between the authorities of a major university who fully appreciated the value of biodiversity, and a group of farmers who fully appreciated the value of mixed and durable agroforestry systems. The main constraint on reaching a sound agreement between the two sides was the legacy of confused land tenure and erratic management over the previous several decades, which made it hard to develop a common sense of purpose in trying to solve the problem.

In Hainan, the main problem was that government had committed itself to a wide range of major development projects which had been planned or implemented without taking biodiversity or any other aspect of the environment into account. The environment was thus considered too late in the island's development to allow the preservation of more than a few fragments of natural ecosystem, and even these would require exceptional efforts to save. The arrangements proposed were innovative in the Chinese context, since they relied on planning with the participation of all groups affected in each area even though they retained a top–down orientation for political reasons. The proposals were well received by the Provincial Government and the ADB, and follow-on work was expected to implement at least some of the projects. The area where progress was most rapid was Jianfengling, and a National Forest Park was constituted there in September 1992, and all logging in Jianfengling's natural forests was banned from January 1993 (Holdgate, 1993). The ADB has since agreed to finance the conservation of the Jianfengling area (M. W. Holdgate, personal communication, February 1994).

In the early 1990s, there was greater scope in the Philippines than in China to use unofficial means to involve local people in solving conservation problems. This is partly because of rapid change in the political and administrative context of conservation in the Philippines at that time, especially due to the Local Government Code of 1991 which came into effect in mid-1992 (MacDonald, 1992; Brillantes, 1993; Nolledo, 1993). The Code transferred certain responsibilities from central government to the various levels of local government unit (LGU, i.e. the province or city, municipality and barangay), while also encouraging NGOs to take an active role in developing local autonomy. The areas of responsibility which the Code sought to devolve included most services in the fields of agriculture, public works, social welfare and health. Others affected include school building programmes and community-based forestry projects, such as social forestry and management of community forests up to 50 km^2 in area.

Meanwhile, certain regulatory powers were also devolved, including those governing the reclassification of agricultural land and the enforcement of fisheries and environmental laws. No reform of this magnitude could be attempted quickly without some difficulties, as people adjusted to their new roles and reacted to new challenges and opportunities. It proved hard, for example, for LGUs to absorb government personnel with certain technical skills, such as in environmental management (Brillantes, 1993). The ability of the LGUs to raise revenues themselves, and to share in national revenues, was greatly strengthened. In fields affected by decentralization, though, the LGUs themselves were required to make budget allocations to support tasks (such as environmental management) in which few of the individuals concerned had been trained.

On the other hand, the process of decentralization was consistent with the efforts of environmental NGOs. This was because individual activists could exert greater influence than formerly, within much smaller and more autonomous geographical and social units. Local decision-making processes became at least potentially more transparent, and links between decisions and their environmental consequences became tighter and more obvious to all. This implied that the environment was more likely to be seen as a local asset, which individuals could understand and influence, rather than as a concern of outsiders, and central government in particular.

Proposals to engage all stakeholders in the future of Mt Makiling's forests within a new NGO were thus consistent with the Philippines' Local Government Code in 1993, but analogous proposals could not have been tabled credibly in China in 1991. A trend in the Philippines since the late 1980s had been towards the idea of local partnerships to manage natural resources, with government agencies collaborating closely with NGOs and farmers' groups to protect forest resources. The design of the Makiling project reflected this approach, since it advocated a formal partnership between UPLB as the official management agency, and certain groups of local farmers. Their interest in collaboration was to be strengthened through the reform of land tenure, and by education and participation in development planning. Both sides would also benefit from a more systematic attempt to capture financial and other benefits from the Reserve as a reward for protecting it.

The project design deliberately made clear, however, that not everyone using the Makiling forest could be accommodated there, and not all forms of environmental usage were to be tolerated. In a small, valuable forest, there was no room for new settlement by people from far away, nor for farming by temporary tenants who were not party to comprehensive and socially policed agreements on their use of land and forest resources. In other words, nonresident farmers and new arrivals would have to be moved on if the Reserve's natural ecosystems were to survive. This was an essential part of a whole policy, the rest of which was focussed on building a long-term partnership between resident farmers and permanent institutional stakeholders.

Resolution of the main conflict on Mt Makiling was well underway by 1994 (P. Guy, personal communication, March 1994). Resident farmers had been encouraged to demarcate their own farms within the Reserve, and had largely done so in collaboration with the University, in accordance with contracts allowing them to stay subject to eviction if additional land was cleared. Meanwhile, committees had been established by the farmers to police their use of the Reserve in accordance with their agreements with the University. These committees had been officially deputized by the Secretary of the Department of Environment and Natural Resources, and had begun to report (and in some cases to punish) illegal activity on the mountain.

The Makiling project illustrates the potential complexity even of a local conservation problem affecting a small forest reserve, in this case with many themes needing to be woven together into a comprehensive project design. These themes include both on-site and off-site biodiversity management, inventory and use of biodiversity information, institutional reorganization, spatial planning, participation and transparency, and education ranging from public and primary to advanced university levels.

6

Costa Rican linkage projects

6.1 Introduction

International awareness of biodiversity in Costa Rica was sharply increased during 1990–91 by a succession of articles in the technical literature (Eisner, 1990; Tangley, 1990; Gámez, 1991a, 1991b; Janzen, 1991; Sandlund, 1991), popular science journals (Aldhous, 1991; Joyce, 1991; Wille, 1991), and newspapers (Lyons, 1991). The immediate focus of much of this attention was the Costa Rican National Biodiversity Institute (INBio), which had embarked on an ambitious inventory project aiming to document all Costa Rica's 500 000 or so native species. This was especially newsworthy at a time when the world's governments were beginning to negotiate the Convention on Biological Diversity. A startling additional factor was that INBio had just signed a multi-million dollar biodiversity prospecting contract with Merck & Co., the world's largest pharmaceutical corporation. The aim of this was to establish terms for INBio and Merck to collaborate in finding and developing commercial products from amongst the country's wild species.

It soon became clear that the biodiversity inventory and prospecting contract were only two aspects of a much more comprehensive response to conservation issues in Costa Rica. Presentations by scientists linked to INBio, such as that by Daniel Janzen in London in October 1991, showed that the inventory process was spinning off a variety of new concepts and techniques. Rural people were working as parataxonomists, and tens of thousands of specimens were being processed at INBio and labelled with unique, computer-legible bar codes. Meanwhile, data on biodiversity were being managed within interactive databases, which had been specially designed both to serve the needs of many different users and to safeguard Costa Rica's national interests. Moreover, there was a process underway involving general reform of the country's protected-area system, empowerment of local people in man-

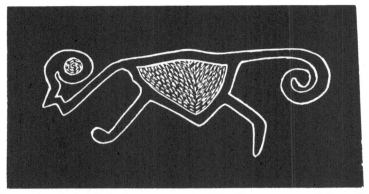

Spider monkey design by indigenous people of the Brunca region of Costa Rica.

aging the reserves, and pioneering work on new ways of financing conservation sustainably. The total package as advertized was quite spellbinding.

A general invitation was extended by INBio for interested people to visit Costa Rica to see for themselves. This was accepted by Shirley Conover at Dalhousie University in Canada as a reason to develop links between Costa Rica and the EMDI and ERMP projects which Dalhousie was managing in Indonesia and the Philippines respectively (see below). By late 1991, both projects had reached a stage at which attention needed to be paid to biodiversity management issues, and Costa Rica was clearly becoming an influential model in this field. It was therefore decided to begin an informal linkage project, in the first instance between Indonesia and Costa Rica. The aim was to assess the Costa Rican biodiversity management strategy, and to see what might be learned and perhaps later applied in Indonesia. Before describing the process of communication between biodiversity managers in these two countries, and later in others, the circumstances of Costa Rica should be explained in more detail.

6.2 Conservation issues in Costa Rica

6.2.1 Land and biodiversity

Costa Rica is located in Central America between Nicaragua and Panamá (Fig. 6.1), and is therefore a place where North American communities of plants and animals meet and overlap with South American ones (Rich &

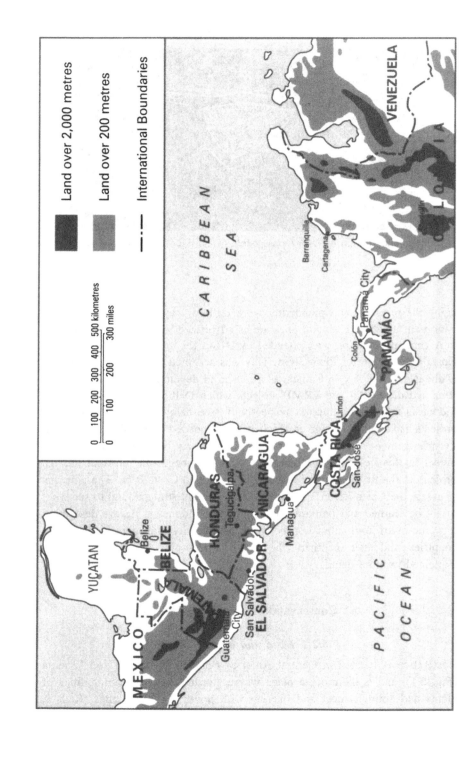

Rich, 1983). The topography and geology are diverse, with both volcanic and sedimentary areas and mountain ranges running north to south (Castillo-Muñoz, 1983). The climate is also varied, due to easterly trade winds and the rain shadows produced when they meet the mountains, while total rainfall each year varies locally from 2500 mm to 5000 mm (Coen, 1983).

These factors contribute to a wealth of different ecosystems, ranging from evergreen rainforests on the Caribbean side of the country to dry seasonal forests on the Pacific side (Hartshorn, 1983a). Costa Rica is therefore very rich in wild species, with up to 13 000 higher plants, 10 000 fungi, 205 mammals, 848 birds, 214 reptiles, and 162 amphibians (WCMC, 1992; Gámez *et al.*, 1993). Most of the invertebrate fauna is poorly known, but there are thought to be about 365 000 species of arthropods and 85 000 other invertebrates (Janzen, 1991). The total number of native species is thus thought to be close to 500 000 species.

6.2.2 Origins of a biodiversity management strategy

Most of Costa Rica's 3 million or so people live in the Valle Central around the capital San José, where the economy is based on farming, light industry, and tourism, although elsewhere in the country plantations and cattle ranches are widespread. Most of the 51 000 km^2 land area is privately owned, and since government has little power to regulate use of private land, its owners were free to invest in forest clearance. During the 1970s and 1980s the country had one of the world's highest rates of deforestation, with almost 4% of forest land cleared annually in 1981–5, a rate similar to that in Nigeria in those years (Hartshorn, 1983b; Repetto, 1988, 1991). This convinced local conservationists that virtually all private lands were destined to be cleared of natural ecosystems, and that if most of the country's wild species were to survive, they would have to do so in nature reserves under public ownership (Gámez *et al.*, 1993).

The strategic challenge was therefore to secure representative and viable samples of all of the country's natural ecosystems, within reserves owned by government and securely managed in the public interest. To achieve this, several problems had to be overcome. Firstly, although about a quarter of

Fig. 6.1 Central America and Costa Rica

Costa Rica's land area had been acquired by government over the years, the various protected wildlands were administered as different kinds of reserve by a number of different agencies. The main legislation was the Forestry Law, but this placed the National Parks Service in charge of National Parks and Biological Reserves, the Forestry Directorate in charge of Forest Reserves and Protection Zones, and the Wildlife Service in charge of Wildlife Refuges. Although all were under the authority of the Ministry of Natural Resources, Energy and Mines (MIRENEM), the Forestry Law tended to prevent them being managed in a co-ordinated way. There were also other categories of land area, such as Indigenous People's Reserves, which were potentially important for conservation but which were controlled by other agencies.

Management of the national reserve system was therefore in urgent need of being consolidated and rationalized. This might have implied centralizing control, but there were two main constraints on doing so, one political and the other financial. Costa Rica has a relatively egalitarian distribution of wealth and well-established democratic institutions and processes, and much of the electorate is distrustful of excessive concentrations of power. Moreover, by the late 1980s Costa Rica was deeply in debt, and there was a political consensus in favour of reducing rather than increasing the size of government and the public-sector wage bill. Proposals to create a new, centralized bureaucracy to run the reserve system were therefore unlikely to be welcomed by the nation's decision makers, electorate or creditors.

Another factor was that, even if the administration of the reserve system could be rationalized, the reserves themselves were not adequate to preserve biodiversity and ecological services. Many of them were too small, and others had private lands interposed between areas of protected natural habitat. To solve this problem, government would have to acquire more land, but was legally unable to expropriate private land without paying full market value for it. This meant, in effect, that conservationists were encouraging government to buy land, at a time when the national treasury was already stressed by its high levels of external debt.

None of this would have resulted in action if there had not been a strong conservation lobby in the country. During the 1980s, however, certain Costa Rican biologists moved into positions where they could influence government to improve the national reserve system. An early effect was to create a government-endorsed NGO, the National Parks Foundation (FPN), the role of which was to support an international fund-raising campaign for Costa Rica's reserves, and to buy land on behalf of government. From 1987 onwards, the FPN also facilitated a series of debt-for-nature swaps, by which conservation groups bought Costa Rican debt at a discount on the international secondary market. This debt paper was then donated back to the Central Bank in return for Monetary Stabilization Bonds, the interest on which was to be used to finance conservation projects (WCMC, 1992).

This arrangement required influential assistance at high levels in government, and the role of the conservation lobby became particularly important in 1986–90, when the National Liberation Party (PLN or *Liberación*) held both the Presidency and a majority in the Legislative Assembly. In this period, there was a growing realization that solving the structural problems of the country's reserve system would require a more complete effort than had previously been contemplated. A National Biodiversity Planning Commission was therefore appointed by presidential decree on the 5th June 1989. It had 9 members, representing 3 ministries, the National Museum, 2 universities, the National Scientific Research Council, and two NGOs (Gámez *et al.*, 1993). Its role was to design a comprehensive strategy for conservation in Costa Rica.

6.2.3 *The National Biodiversity Planning Commission*

The Commission started its work from the assumption that biodiversity was economically valuable, and should therefore be preserved and used for the public benefit (D. H. Janzen, personal communication, February 1992; R. Gámez, personal communication, March 1993). Saving biodiversity was to be achieved primarily in the wild, by maintaining large blocks of natural forest under public ownership. From government's previous experience in acquiring and managing land for conservation, the Commission realized that protecting natural forests would be expensive in financial terms, regardless of its economic justification. Conservation would thus generate a need both for capital and for secure recurrent financing.

The Commission also realized that forest protection would in the end be unsuccessful unless the people living around each protected area were willing to help protect it. Thus, the permanence of boundaries between protected and unprotected land would ultimately depend on the public appreciating the benefits of retaining public lands under natural forests. It was assumed that there would always be an incentive for land-hungry individuals or groups to seek access to 'unused' public land, which would need to be countered by a clear and durable social consensus in favour of maintaining natural forests. In the long term, it was hoped that education would sustain this consensus by helping people to appreciate the economic, intellectual, and aesthetic values of forests and biodiversity. Meanwhile, however, there would be a need to find ways for intact forests to generate financial rewards for people who had influence over the future of each forest area.

The Commission therefore considered it important for people living around protected areas to receive both education about biodiversity and also direct benefits from preserving it, in the form of employment and income. It felt that government and local communities both have a claim on revenues generated by protected areas. Government claims would arise because of public investment in protection, while community claims would arise because voting support is essential to the security of those investments. The Commission therefore sought ways for each of Costa Rica's reserves to yield sustainable revenues, which could be used to reward everyone concerned for helping to protect them.

The Commission recommended a strategy to reconcile these various factors. Firstly, it proposed new legislation to consolidate a National System of Conservation Areas (SINAC). The various reserves were to be assembled out of existing government and private land-holdings, to create a number of areas each with the common management objective of preserving biodiversity. These were to be co-ordinated and assisted by a central agency (initially the National Parks Service, SPN), but real management authority was to be decentralized and devolved to each area, thus helping it to be seen locally as a local asset rather than as a project of central government. Secondly, to help pay for the administration of SINAC and its various component reserves, the Commission recommended a national biodiversity inventory, to find out exactly what made up Costa Rica's biological richness and what it might be used for. It also proposed the creation of a National Biodiversity Institute (INBio) to manage the inventory in the public interest.

The Commission was set up by a *Liberación* president and government, but that administration was displaced in the February 1990 election by the more conservative Christian Social Unity Party (PUSC, or *Unidad*). This delayed the implementation of some of the Commission's recommendations, since the new government was less sympathetic to the idea of decentralization than its predecessor. Although INBio came into being before the changeover, and legislation to create SINAC was approved unanimously at the committee stage, the new law was not passed by a full vote of the Legislative Assembly (R. Gámez, personal communication, October 1993). This delayed the transfer of authority from San José to the Conservation Areas. By late 1993, their legal position was uncertain, since several had followed the SINAC model in localizing administrative and financial arrangements, and some had also received external assistance conditional on a law which had not yet been passed. The election of February 1994, however, returned *Liberación* to power, with a mandate to renew the process of constituting SINAC more or less as envisioned by the Commission, and to give priority to biodiversity management and ecotourism (Burnie, 1994; A. M. Piza, personal communication, April 1994).

6.2.4 The National System of Conservation Areas

Because of the changes of government in the period 1989–94, implementation of the SINAC concept was patchy and confused. The main components of SINAC were agreed by 1992, however, since they were based on the country's existing reserves. They comprised the eight areas known as Amistad, Arenal, Cordillera Volcanica Central, Guanacaste, Osa, Tempisque, Pacífico Central (Isla del Coco), and Tortuguero (Fig. 6.2). The aim of SINAC legislation was to consolidate the various kinds of protected wildland in each of these locations, and to manage each as an ecological unit. Economic links between these areas and the people living around them were to be made stronger, and those people were to be actively involved in managing them. To ensure local participation and the local capture of economic benefits, responsibility for managing reserve budgets and staff was to be devolved from the national capital to the areas themselves.

The SINAC concept has been applied most fully in the Guanacaste Conservation Area, which was formed in 1989 from three national parks, an experimental forestry station, a recreation area, and private lands, making a total of about 1100 km^2 of dry forest stretching inland from the northern Pacific coast (Janzen, 1983, 1986b, 1988a, 1988b, 1991; Fig. 6.3). An endowment fund was established for the area, and by early 1992 this had reached about US$11 million. The reserve's policy called for its annual yield to stabilize in real terms at a level consistently equal to or greater than the area's annual budget (D. H. Janzen, personal communication, February 1992). This budget supported the eight management programmes of fire control, security, sector supervision, biological education, extension, research, ecotourism, and ecological restoration (S. Marin, personal communication, March 1993). There were 84 employees in 1992, of whom nearly two-thirds were indigenous to Guanacaste, and almost one-third were women. The area included four research stations, where parataxonomists were trained, and worked on the biodiversity inventory (see below).

The management programmes and staff related to a central Directorate, which was divided into Sub-Directorates for Operations and Management and for Eco-Development. All were supported by a single accounting and

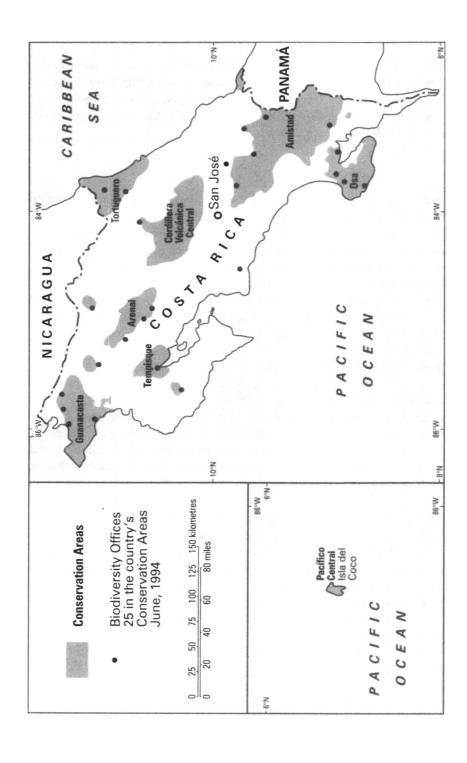

Conservation Areas

- Biodiversity Offices 25 in the country's Conservation Areas June, 1994

0	25	50	75	100	125	150 kilometres
0	20	40	60	80 miles

CARIBBEAN SEA

NICARAGUA

Guanacaste

Tortuguero

Arenal

Tempisque

Cordillera
Volcánica
Central

○San José

COSTA RICA

Amistad

Osa

PANAMÁ

PACIFIC OCEAN

10°N

10°N

8°N

86°W

84°W

86°W

84°W

6°N

Pacífico
Central
Isla del
Coco

PACIFIC OCEAN

6°N

6°N

86°W

86°W

personnel office directly under the Director. A Programme Chief was responsible for each management programme, which was designed according to clear budget ceilings. Because of their long-term budgetary significance, new staff appointments within any one programme required the approval of all eight Programme Chiefs and the Directorate, the heads of which together made up the Technical Committee. This was one of three committees which supported the Director in running Guanacaste, the others being the Research and Alterations Committee, and the Regional Committee.

The Regional Committee was the main guarantor of local control of Guanacaste. In late 1993, it comprised 35 selected individuals representing 47 local institutions, communities, and corporations, who elected an 11-member Board of Directors. The Committee had the authority to approve Guanacaste's annual workplan before it was submitted to MIRENEM in San José. The Committee was deliberately kept closely involved in decisions affecting the Conservation Area, and was able to veto proposals by other groups where these were considered likely to have an unacceptable impact on it.

The Guanacaste Conservation Area is an important model of local management, sustainable cost recovery, and conflict resolution. It demonstrates a number of important principles, although not without some degree of controversy because, by late 1993, the administration of Guanacaste had gone far beyond existing law and was, in effect, waiting for the law to catch up. There were also reservations in some quarters about increasing the influence of local NGOs, about the loss of central authority implied by decentralization, and about the success of Guanacaste in obtaining financial resources compared with that of other areas. There was a feeling that such resources should be spread more evenly through the national conservation system, and that central government should retain the power to ensure that this was done.

If Guanacaste is perhaps the most advanced and best-funded conservation area in Costa Rica, contrasting examples can be found in the Brunca Region, where uncertainty over legislation and political leadership have had a more severe effect. This is a 9500 km² area in the south of the country, inland from the Pacific coast to the Continental Divide, and between southern San José Province and the Panamá frontier (Fig. 6.4). It is sparsely populated and relatively poor, and is dominated inland by the Cordillera de Talamanca, which has 17 peaks higher than 2600 m and a maximum height of 3820 m

Fig. 6.2 Costa Rican conservation areas

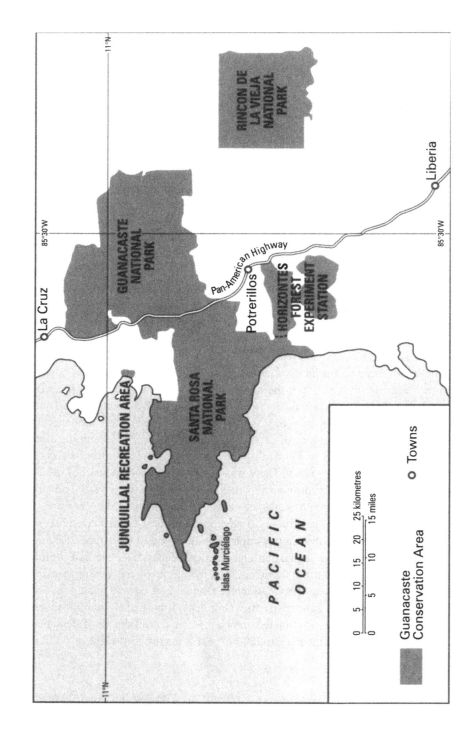

RINCON DE LA VIEJA NATIONAL PARK

GUANACASTE NATIONAL PARK

Pan-American Highway

Potrerillos

HORIZONTES FOREST EXPERIMENT STATION

La Cruz

JUNQUILLAL RECREATION AREA

SANTA ROSA NATIONAL PARK

Islas Murciélago

PACIFIC OCEAN

Liberia

11°N

85°30'W

11°N

85°30'W

| 0 | 5 | 10 | 15 | 20 | 25 kilometres |
| 0 | | 5 | 10 | 15 miles |

Guanacaste Conservation Area

o Towns

at the crest of Mount Chirripó. By late 1993, villages had been established up to over 1000 m on the Cordillera, but forest clearance extended in places up to 1000 m higher than that. The Pacific-facing slopes of the Cordillera are steep, and there was intact forest cover at high altitudes to the north-west and north-east of the Brunca, but widespread forest damage and fire-maintained grasslands elsewhere.

Much of the Cordillera lies within La Amistad International Park, which extends into Panamá. The Amistad Conservation Area (ACA), according to the SINAC concept, comprises the Tapantí, Chirripó and Cahuita National Parks, La Amistad International Park, and the Hitoy Cerere Biological Reserve. As a component of SINAC, Amistad would also include several indigenous people's reserves, forest reserves, and protected zones. These were all included within the Costa Rican part of the Amistad Biosphere Reserve, which extended along the Cordillera into Panamá. The Cordillera de Talamanca is an important bird endemism area, with 51 restricted-range species confined to it (ICBP, 1992, 1995), and it is also a centre of plant diversity (WCMC, 1992; WWF & IUCN, 1995).

An example of the state of the environment on part of the Pacific slope of the Cordillera in late 1993 was provided by the view from the small village of San Jerónimo de Unión (formerly Esperanzas), at about 1200 m elevation. Steep to very steep lands uphill in all directions were extensively deforested, with forest cover persisting only close to the ridge crests, and with signs of fire damage visible in the distance well within the boundaries of Chirripó National Park. Although few areas of active soil erosion were visible, there had been widespread clearance of forest on steep land, and its replacement by rough pasture grazed by cattle and horses. Wherever the slope and residual fertility made it feasible, lands had been planted with coffee, or with annual food crops closer to the village. Pressure on the surviving natural forests was mainly being maintained by fire incidental to pasture maintenance.

The southern part of the Brunca Region is dominated by the Golfo Dulce and the Peninsula de Osa, around which are the intact forests of the Corcovado National Park and various forest and indigenous people's reserves, including the mangroves of the Valle de Diquis, and the Golfito Wildlife Refuge. The Osa Conservation Area will, with SINAC legislation, cover most of the Osa Peninsula and the lands surrounding the upper Golfo Dulce. This

Fig. 6.3 Guanacaste Conservation Area

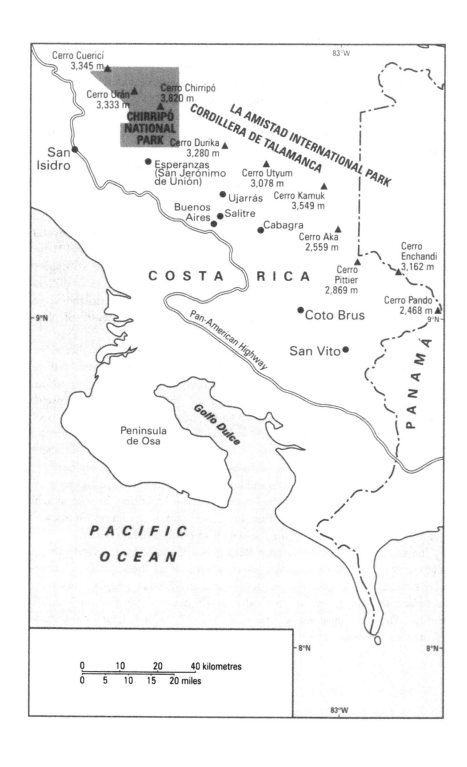

region is another bird endemism area, with 12 restricted-range species confined to it (ICBP, 1992, 1995), while the Osa Peninsula is also a centre of plant diversity (WWF & IUCN, 1995). The Brunca Region therefore extends from one extremely important biodiversity area (the Cordillera) to another (the Gulf and Peninsula), and includes a wide variety of lands and human settlements in between.

Neither the Amistad nor the Osa Conservation Areas had well-established local institutions in late 1993, and both were short of resources. The embryo of SINAC-style management existed at Amistad, since its environmental education programme was developed locally, with SPN in San José merely being kept advised through annual workplans (C. Fernandez, personal communication, October 1993). Some support had been provided to Amistad by the MacArthur Foundation and the Swedish government, and other funds had been allocated to Amistad and Osa through the Global Environment Facility. This support was mainly for infrastructure and human resource development, although by late 1993 a lack of current resources meant that staff and operational cutbacks were imminent. At that time, the ACA was estimated to have about 40% of the staff levels needed to resist pressures on the area (C. Fernandez, personal communication, October 1993). In all cases, these allocations were made on the assumption that SINAC legislation would be passed by the Legislative Assembly.

6.2.5 The National Biodiversity Institute

INBio was established to study and use the biodiversity represented by Costa Rica's half-million or so native species. It was constituted on the 24th October 1989 as a private, non-profit, public-interest corporation (Gámez *et al.*, 1993). It is essentially a data-management agency, which exists to serve the needs of people who wish to use information related to biodiversity (Janzen, 1991, 1992a; Gámez, 1991a, 1991b; Sandlund, 1991; Gámez *et al.*, 1993). It has four divisions which are responsible for different aspects of its task.

The inventory division aims to document all the species occurring in all of the country's reserves. Samples of all species are collected, stored, described, and identified, and records made of their locations, distribution,

Fig. 6.4 The Brunca Region

natural history, ecology, morphology, behaviour, phenology, and genetic variation. The information management division is responsible for managing the enormous quantities of data generated by the inventory, in a way which matches the needs of potential users. The prospecting division works with private companies to identify materials with commercial potential, and to assist in their commercial development. Lastly, the information dissemination division manages non-commercial information as a service to education, science, conservation, tourism and other sectors with a need for data of this kind.

By late 1993, INBio possessed a well-established and rapidly growing infrastructure in Heredia, near San José, and was running 28 biodiversity stations in all ecological zones of the country. These were the field sites of the national biodiversity inventory, which was then yielding specimens for processing by INBio at a rate of about 900 000 each year. In the first years of the inventory, there was a strong emphasis on plants and insects, reflecting both the availability of local expertise and the extent of missing knowledge. About 15% of the national flora and 85% of the insect fauna were estimated to be unknown. The inventory was due in 1994 to include spiders, molluscs and tunicates (E. Sancho, personal communication, October 1993).

The activities of INBio are significant for Costa Rica's nature reserves in a number of ways. Firstly, the inventory was designed partly with a view to the development of commercial products, in response to the long-term financial needs of SINAC. An important aim was for Costa Rica to obtain a share of any resulting revenues, in order to help pay for national conservation efforts. INBio therefore seeks to ensure that all biodiversity research occurs under legal contracts, which guarantee local returns on commercial products (Janzen, Hallwachs, Gámez, Sittenfeld, & Jiminez, 1993; Gollin, 1993).

INBio also has a more direct role in biodiversity prospecting, since it prepares catalogues of known species which can be used to rent access to selected species for limited periods (Sittenfeld & Gámez, 1993). Samples may then be provided by INBio for use according to prospecting contracts, which provide for advance payments and royalties (Barbier & Aylward, 1992; Laird, 1993; Coghlan, 1994). Advance payments are often in the form of a 10% surcharge payable to the central account of the Conservation Area system. Royalties are also payable on commercial applications of products provided by INBio, and are divided equally between INBio and the reserve system.

The biodiversity inventory has a role in conservation which goes well beyond fund-raising, since INBio encourages local people from around the nature reserves to participate fully in the inventory. These are selected, trained, and employed as 'parataxonomists', who undertake the initial collecting and processing of specimens which are then sent to INBio for final identification and analysis (Janzen, 1992a; Janzen, Hallwachs, Jiminez, & Gámez, 1993). The parataxonomists live and work in the area where they were recruited, and they have an important role in communicating their special knowledge and enthusiasm to the rural people around their homes, thereby teaching them about biodiversity and conservation. In this sense, their role in public education is similar to that of the village liaison assistants used around Cross River National Park in Nigeria (Chapter 3).

The original reason for using parataxonomists was not related to environmental education, and it was instead a pragmatic response to the scale of the task facing the national inventory. INBio's managers estimated that to use only the qualified taxonomists available in Costa Rica would simply take too long, and it was decided to use parataxonomists to help speed up the process. They have allowed a great saving in expert staff time, and the inventory has proceeded rapidly with several million specimens having been received by INBio. These have been processed by INBio's curators under guidance by national and international experts, the latter often working from personal interest with independent financing.

The management of biodiversity data is central to INBio's role, making the design of a suitable database extremely important. By 1992, INBio had defined the needs of its users sufficiently clearly to have begun developing a system able to keep very large amounts of alphanumeric (specimens, species and literature), graphical (pictorial) and geographic (mapped) data in separate but interactive fields. An agreement between INBio and the Intergraph Corporation allowed INBio to obtain a powerful ORACLE-UNIX system and to collaborate in developing the necessary software (Gámez *et al.*, 1993). This system was intended to meet the needs of users ranging from scientists to government planners, school children, and private businesses.

6.3 Linkage projects

6.3.1 Indonesia and Costa Rica

The Ministry of State for Population and Environment (KLH) was the central government agency responsible for coordinating and developing policy on population and environmental management in Indonesia. Until 1993, when it was divided into two new ministries, an important part of the role of KLH and its long-time minister, Dr Emil Salim, was to analyze environmental issues and make them more relevant and accessible to other ministries. It also sought to build consensus within the Indonesian establishment around legislation and procedures to support sustainable economic development. Achievements of KLH included creating the Environmental Impact Management Agency (BAPEDAL) in 1990, and formulating laws such as Act 24 of 1992 and Government Regulation No. 51 of 1993 to govern spatial planning and environmental impact analysis respectively (Dick & Bailey, 1992; BAPEDAL, 1993, 1994; Dahuri, 1994).

The Government of Canada had assisted KLH since 1983 through a joint venture between KLH and Dalhousie University known as the Environmental Management Development in Indonesia (EMDI) Project. This had three Phases: EMDI-1 in 1983–6; EMDI-2 in 1986–9; and EMDI-3 in 1989–94 (Miles *et al.*, 1993). In 1991, KLH was responsible for co-ordinating Indonesia's role in negotiating Agenda 21 and the Conventions on Biological Diversity and Climate Change, all of which were due to be signed at the United Nations Conference on Environment and Development (UNCED) in 1992. Because of this, a decision was made at KLH to study international practice in biodiversity management, so as to assess Indonesia's existing policy and to suggest improvements where necessary. These efforts were assisted by EMDI, and began with an EMDI–KLH technical mission to Costa Rica in early 1992, during which Guanacaste Conservation Area and INBio were both visited (Caldecott & Alikodra, 1992).

Two subjects were of special interest to Indonesia, the first of which was an urgent need to find affordable and cost-effective ways to manage the country's vast system of nature reserves (Chapters 4 & 7). Secondly, there was interest in finding ways to make an effective inventory of all the reserves and the biodiversity they contained, and then to manage the resulting data in a way that would contribute both to conservation and to national development. The preliminary Indonesian assessment of work in Costa Rica concluded that: 'The Costa Rican biodiversity management strategy is impressively comprehensive, theoretically sound, consistent with worldwide conservation experience, and practical in its methods and objectives. By decentralizing management of the [protected area] system, the Conservation Areas themselves are treated as being critical units for biodiversity management, and opportunities have been created for this management to be locally appropriate, to involve and be supported by local populations, and ultimately to create a financially self-sufficient conservation entity. The creation of INBio, meanwhile, provides a way eventually to relieve Government of the financial burden of maintaining the Conservation Area system, and to derive wealth more generally from publicly-held biodiversity assets, while justifying the retention of intact ecosystems which will sustain a wide range of private-sector investments and global service functions indefinitely' (Caldecott & Alikodra, 1992:11).

It was recommended that further links be developed between Costa Rica and Indonesia, and the Minister of KLH invited the Director General of INBio, Dr Rodrigo Gámez, to visit Jakarta in October and November 1992. This visit included meetings with senior officials of all the key agencies involved in biodiversity management in Indonesia, including the Indonesian Institute of Sciences (LIPI) and the Directorate-General of Forest Protection and Nature Conservation (PHPA). A formal statement of 'Cooperation and Shared Ideals' was signed on the 30th October 1992 between the Chairman of LIPI and the Director General of INBio, and witnessed by the minister of KLH. This was followed by a large meeting at which Dr Gámez described INBio and Costa Rica's biodiversity programme to over 100 invited guests representing a wide range of Indonesian government agencies, the Indonesian private sector, and both governmental and non-governmental donors.

In discussions during the meeting, it was agreed that there was a need for a co-ordinated response by Indonesia. This would involve KLH seeking a consensus on the need for a National Biodiversity Commission to examine relevant issues, as called for in the *Biodiversity Action Plan for Indonesia* (BAPPENAS, 1991). Consensus was also to be sought for a steering committee to advise LIPI and PHPA in jointly creating an Indonesian Biodiversity Foundation to act in the public interest as a broker in linking conservation, science, investment and sustainable development. At the same time, consensus was to be sought on the choice of site for a pilot project through which to explore the issues and their practical application in Indonesia. It was also agreed that there was a need for an exchange programme between Indonesia and Costa Rica. This was to involve appointment of international liaison staff at LIPI, PHPA, and INBio, and the sending of selected staff from LIPI, PHPA, KLH, and other Indonesian agencies to Costa Rica, and reciprocal visits by staff of INBio and MIRENEM to Indonesia. Such exchange visits were to lay the groundwork for a mutual learning process to the benefit of both countries.

Lastly, it was agreed that there was a need for the active support of the international donor community, with the aim of speeding up the development of collaboration between the two countries, and helping to initiate a process of south–south co-operation in the spirit of the Convention on Biological Diversity. The text of a draft proposal for support with which to implement the next phase of the exchange programme was read into the minutes of the meeting, and was to be developed in further discussions between KLH, LIPI,

PHPA, INBio, and interested donors. The World Bank representative at the meeting indicated that requests for help with which to develop the project would be received sympathetically, and this was echoed later in informal circumstances by the Ambassadors of the United States and Sweden, the Resident Representatives of the World Bank and UNESCO, and the Development Counsellors of Canada, the United States, the United Kingdom, and the Commission of the European Communities.

6.3.2 The Philippines and Costa Rica

As of 1993, the University of the Philippines at Los Baños (UPLB) was custodian of an important biodiversity asset, the Makiling Forest Reserve (Chapter 5), and at that time UPLB was seeking ways to save this reserve in the face of severe pressure on land from immigrant and resident farmers. In doing so, UPLB was trying to develop solutions to a class of resource-management conflicts which are widespread in the Philippines. As a leading university, UPLB hoped to exert influence at a national level by saving, studying, and using biodiversity in a way consistent with the national priority for community participation in resource management.

The Canadian Government had been assisting UPLB since the early 1990s through the Environment and Resource Management Project (ERMP), which was a joint undertaking of Dalhousie University and UPLB's Institute of Environmental Science and Management. As part of a study of biodiversity management options for UPLB, Dalhousie initiated a technical mission by ERMP and UPLB to Costa Rica (Caldecott, 1993a). During this, the team visited the Guanacaste Conservation Area and INBio headquarters, and a joint statement on co-operation was signed on the 30th March 1993 between the Dean of the College of Forestry of UPLB and the Director General of INBio. This envisioned regular communication, exchange visits, joint training, and consultations on helping their respective governments to protect one another's property rights within the framework of the Convention on Biological Diversity.

Several areas of special interest to UPLB emerged from discussions at INBio. First was the concept of a comprehensive biodiversity management programme, taking into account new ideas on complete, user-oriented biodiversity inventories and use of common standards for collection and management of information. Second was the idea that such inventories could help to increase community awareness of biological issues and to enhance local participation, especially by involving parataxonomists. Third was the need to design an inventory with multiple complementary aims, ranging from basic taxonomy and ecology of each species to the idea of using that information for both commercial and non-commercial ends. Fourth was a reaction to the urgency of conservation problems on Mt Makiling, putting a premium on negotiating appropriate research agreements, while also improving institutional support for biodiversity management in the Philippines. Lastly, UPLB became aware of the potential value of further contact with INBio, but also that this would impose demands on INBio's resources. It was therefore agreed that UPLB would try to help INBio obtain additional support so that it could continue to provide such services to the Philippines and other countries.

6.3.3 Living Earth and Costa Rica

The Living Earth Foundation is a UK-based NGO which works in environmental education, and which by late 1993 had active field projects in Cameroon, Venezuela, and Brazil. These projects had developed techniques for helping rural teachers to use the physical, biological, and social environment of their schools as teaching resources. This approach was aimed at relieving a chronic shortage of teaching materials provided by government, while also strengthening links between schools and the communities where they are located. By also helping teachers to meet and work with one another, several thousand teachers, especially in Cameroon, had organized themselves into support networks (or 'cluster groups') able to deliver cost-effective environmental education to large numbers of people.

Living Earth's philosophy and operational practice was focussed on encouraging rural people (and, in Brazil, the urban poor) to understand, take charge of and use their own environment, largely independently of central government. This implied a degree of convergent thinking with conservationists in Costa Rica, many of whom were concerned with decentralized, participatory management of the environment in and around nature reserves. The involvement of parataxonomists in the national biodiversity inventory was also of interest to Living Earth, because of their powerful educational impact at the village level whence they are recruited and to where they return to work. Another factor was that Living Earth in Cameroon and the Ministry of Public Education in Costa Rica had both been promoting environmental education by radio, and in September 1993 a team of teachers from Living Earth Cameroon visited Costa Rica to share experience in this field.

Living Earth UK then decided to study environmental education in Costa Rica from the perspective of INBio and the management of Conservation Areas, and a team visited the country in September and October 1993

(Caldecott, 1993b). In contrast to earlier visits by teams from Indonesia and the Philippines, Living Earth also sought to explore options for starting a project in Costa Rica, so an aim of the visit was to assess the country's need for assistance in environmental education. This assessment was focussed on the Brunca Region, starting with the Pacific slope of the Cordillera de Talamanca because of Living Earth's interest in collaborating with the existing Amisconde project near the southern boundary of the Chirripó National Park. (Box 6.1). This project was based at San Jerónimo de Unión, the environment of which was described above.

Box 6.1 The Amisconde project in Costa Rica

The Amistad Conservation and Development (Amisconde) project was conceived in 1991 by individuals associated with Conservation International, a US-based NGO, and the Tropical Science Centre in San José. By late 1993, funding had been secured from the McDonald's Corporation, and the project covered San Jerónimo and three nearby settlements in Costa Rica, while also working at another project site near Cerro Punta in Panamá, also close to the Pacific side of La Amistad International Park. The project responded to environmental degradation on steep lands around San Jerónimo through a vigorous campaign to reform agriculture, to prevent forest fires, and to encourage reforestation and soil conservation. It emphasized working with the local Agricultural Producers' Association, the Youth Group and the Women's Group, which have collaborated in programmes concerning agriculture, community development, and forestry respectively.

The agriculture programme sought to diversify cropping systems by adding fruit trees, rehabilitating degraded cattle pastures by planting trees and improved grasses, and through soil conservation. It also assisted the Women's Group by financing a nursery for fruit trees, and by advising on a marketing strategy for agricultural produce and on lease-purchasing vehicles and other large items. The community development programme focussed on developing local management skills, providing community infrastructure and financial services through a revolving credit fund. This fund was set up with a local commercial bank to provide loans to farmers, thereby relieving a shortage of credit in exchange for the opportunity to guide investments towards sustainable and financially sound forms of land use, backed by adequate training of the borrower. The forestry programme emphasized reforestation of degraded areas and the prevention and control of forest fires, mainly by helping the Youth Group to establish a nursery for forest trees, by subsidizing reforestation plantings, and by raising awareness, training and equipping people to prevent, control, and combat fires.

By late 1993, the Amisconde project had demonstrated a number of important strengths, including its clear aims, its choice of location according to those aims, and its thorough and flexible planning with balanced input from outside experts and local people. It also possessed dedicated staff, ample technical expertise and financial resources, and it had made good use of existing local infrastructure. Activities strongly supported sustainability, by strengthening local social institutions, by discouraging dependency, by introducing financial and agricultural skills, and by relying on voluntary adoption of new techniques of land use. The main weakness of the project at that time was that its ultimate purpose of protecting the Chirripó National Park was not being communicated. Concern for biodiversity was also poorly represented in the agricultural programme, which emphasized planting of non-native species, rather than using native species in new crop combinations, and the control of exotics in the area.

Elsewhere in the Brunca Region, by late 1993 a number of local and international NGOs were active in environmental education. CARE International maintained an office in San Isidro, from which it supported work in the nearby Quebradas watershed aimed at reducing pesticide use and encouraging soil conservation measures, tree planting, diversified cropping systems, and income-generating microprojects by women's groups. It also supported several local NGOs, including Fundación Tinamastes, which had opposed logging of watersheds and was working with communities and schools, and Kaneblo, which had emphasized strengthening the cultural identity of aboriginal communities, and mobilizing them to protect their interests against outsiders.

Meanwhile, the Fundación Iriria Tsochok represented aboriginal peoples of the Amistad Biosphere Reserve, and sought to help them preserve the environment of the Cordillera de Talamanca against threats such as plantation development, mining, ranching, fires, and proposed oil pipelines and roads across the Cordillera. The original intention of this group was to restore local participation in the management of the Biosphere Reserve, in response to the failure of the Legislative Assembly to pass SINAC legislation. Most of its initial support came from Conservation International, although other links existed, and there was some tension with government because of conflicts of interest between aboriginal and other groups, and poor communication between those groups.

The Organization for Tropical Studies was also active in the Brunca Region, with a field study centre at the Wilson Botanical Garden at Las Cruces, near San Vito among the foothills of the Pacific slope of the Cordillera de Talamanca. The Botanical Garden was part of the Amistad Biosphere Reserve, and had excellent facilities for visitors and researchers, including interpretative materials and nature trails. By late 1993, however, deforestation had left the Botanical Garden an ecological island. In response to this, an active education programme managed by the Garden had established ecological committees at five communities. The aim of these was to encourage local communities to identify their own problems, and then to maintain a dialogue through which to seek solutions independent of government bureaucracy at the village level.

The Living Earth survey confirmed that the Brunca Region was in need of comprehensive environmental education support, because of its importance for biodiversity and the urgency of threats its environment. The main issue was that activities in the region relevant to conservation and environmental education were rapidly multiplying through the work of local NGOs, international NGOs, governmental donor agencies, and the Costa Rican government. These lacked common or complementary vocabularies, objectives, and standards of training and teaching techniques, while the various groups communicated poorly with each other and often seemed to be in competition.

A proposal to secure additional funds for the Brunca Region therefore aimed to improve training and resources for environmental education in support of all the existing organizations and programmes which were becoming active there. This proposal was evaluated favourably by the British ODA's Joint Funding Scheme, but in mid-1994 Living Earth was still seeking matching support with which to put the project into effect (R. J. Hammond, personal communication, July 1994).

6.4 Discussion

Perhaps the most important lesson of Costa Rican conservation experience is the value of clear and consistently rational thought. Ideas do matter, and problems can be solved if they are addressed in an orderly, flexible, and holistic way, taking into account a wide range of opinions and interests, while being guided by ecological and economic realities. Among these realities are that failing to preserve biodiversity is expensive, and failing to maintain environmental integrity is disastrous. Among the ideas that matter are that environments are most important to the people who live in them, and that those people can become the strongest allies of conservation if they are encouraged and enabled to do so. One effective way to solve complex problems is to use a study commission with a broad-based membership and a strong political mandate, appointed with due regard to potential conflicts of interest and operating in a public and transparent manner.

If these ideas and procedures are well focussed, solutions will emerge within which technology can play a complementary role. Thus, the people who created INBio did not start by visualizing a database of immense complexity and then working out how to fill it up. Instead, they visualized what needed to be done, how to do it, and how a database would be used, before starting to design one. Until that process of thought has been completed, a national biodiversity database will inevitably be the wrong thing to invest in. These remarks are relevant to the fact that a number of donor agencies were seeking to sell biodiversity database systems to Indonesia during 1992, before the Indonesian scientific and government establishments had had a chance to think through their real needs. Indonesian exposure to Costa Rican biodiversity managers had an impact on events to the extent that further discussions occurred between LIPI and INBio at a meeting in Australia in 1993, and that Indonesia also began exploratory discussions with the Intergraph Corporation on obtaining a system similar to that being developed with INBio.

Nevertheless, it is hard to know the actual effect of this contact between Indonesia and Costa Rica. There was, for example, resistance among some Indonesians to the idea that Costa Rican circumstances were relevant to Indonesia. Against this, the Minister of KLH emphasized, during Rodrigo Gámez' visit, that Costa Rica, although small relative to Indonesia, is nevertheless larger than all but five of Indonesia's 17 000 islands. Moreover, the average size of Costa Rica's seven Conservation Areas is about 1700 km², and this is almost the same as the average size of Indonesia's most important large National Parks and Strict Nature Reserves (KLH, 1992). Thus, the scale of the biodiversity units which the two countries have to manage is similar, even though Indonesia has far more of them, and both countries have allocated about one-quarter of their different land areas to conservation purposes.

Even if an important educational event was achieved in Indonesia, there remained the question of how to follow up on it. This applied also to contacts between the Philippines and Costa Rica, although less so in relation to Living Earth, which planned long-term collaboration both within the country and also between Costa Rica and its own projects in Cameroon, Brazil, and Venezuela. International linkages will presumably be more effective the longer they persist and the more regularly they are reinforced. This implies that staff must be assigned in each partner country to nurture the linkage, and that budgets must be allocated to support correspondence, exchange visits, training courses, and mutual attendance at meetings in third-party countries. Some of the most effective exchanges of views can occur between people from countries which are far apart and have little reason to have conflicting interests linked to trade and politics. These same countries, however, are also least likely to have access to any budget line which can be used to support cooperative activities in the field of conservation. This irony conflicts with the urgent need for an international consensus on how best to design and implement conservation projects.

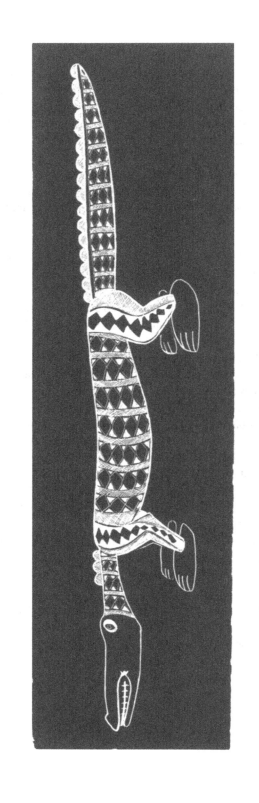

7

Irian Jaya, Indonesian New Guinea

7.1 Conservation issues

7.1.1 The importance of New Guinea

At about 867 000 km², and with its Central Cordillera rising above 4500 m, New Guinea is the largest and highest tropical island in the world. People have lived there for some 40 000 years (Nightingale, 1992), and its diverse aboriginal groups speak as many as 650 distinct languages (Silzer & Heikkinen, 1984; Diamond, 1986; Muller, 1990; Suparlan, 1994). Each of these represents a unique human world view, and a set of cultural features and ethnobiological knowledge developed through long experience of local environments in the homeland of each group.

New Guinea lies between the equator and 12° south, just to the north of Australia (Fig. 7.1), with which it shares a common continental shelf (the Sahul Shelf) and many biogeographic affinities. The latter also extend into the Pacific and throughout Malesia (Gressitt, 1982; Chapter 4). The island contains a vast array of natural ecosystems, ranging from alpine meadows and glaciers in the high mountains to lowland swamp forests and coastal mangroves. It is large enough to contain centres of endemism, and there are many cases of local speciation among birds and mammals (Diamond, 1986).

As a whole, New Guinea contains at least 11 000 species of higher plants, 2000 ferns, 200 mammals, 740 birds and 275 reptiles, and somewhere between 50% and 90% of the species in all these groups are endemic to the island (Petocz, 1984; Collins, Sayer, & Whitmore, 1991). This makes the

Crocodile design by Asmat people of southern Papua.

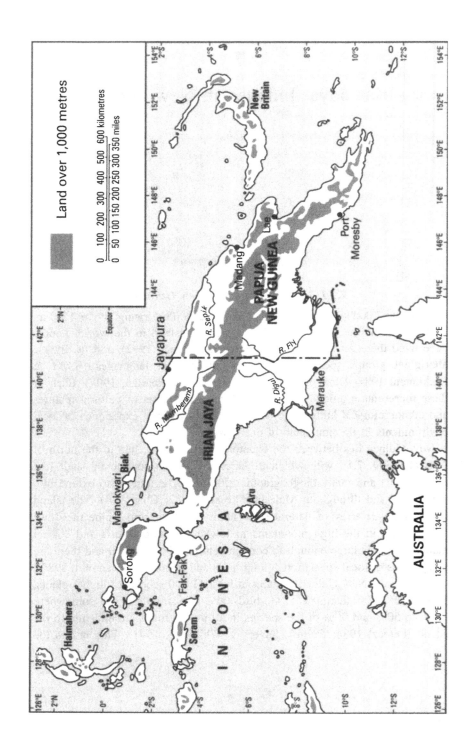

flora and fauna of New Guinea similar in total species-richness and endemism to that of Madagascar, another large (582 000 km²) tropical island which is famous for its biological richness. A difference between New Guinea and Madagascar is that, having been isolated from Africa for up to 165 million years, many of Madagascar's taxa are endemic at a higher level than that of the species (Sayer, Harcourt, & Collins, 1992).

New Guinea is divided almost in half by the frontier between the Indonesian province of Irian Jaya to the west, and Papua New Guinea (PNG) to the east. The border is mostly an arbitrary straight line, running north–south except for a stretch along the Fly River, and the two parts have quite different modern histories and administrative systems (Diamond, 1986). At some 415 000 km² in land area, Irian Jaya is about the same size as California and 20–30% larger than Malaysia or the Philippines. It is the largest of Indonesia's provinces, and comprises the nine *Kabupaten* (Districts or Regencies) of Jayapura, Jayawijaya, Merauke, Biak-Numfor, Yapen-Waropen, Paniai, Fak-Fak, Manokwari, and Sorong (Fig. 7.2).

Irian Jaya is an almost 50% sample of New Guinea and, like the whole island, it contains scores of ecosystem types and sub-types. Maps prepared in the late 1980s showed that about 86% of its land area was covered by natural ecosystems (RePPProT, 1990), which were classified into the following kinds:

- coastal forest (3800 km²);
- tidal forest (21 900 km²);
- peat swamp forest (22 600 km²);
- fresh-water swamp forest (33 100 km²);
- lowland rain forest (176 800 km²);
- heath forest (7100 km²);
- forest on limestone and ultrabasic rocks (37 900 km²);
- lower montane rain forest (22 100 km²);
- upper montane rain forest (22 400 km²); and
- monsoon forest (11 000 km²).

Irian Jaya has about 639 bird species, 154 mammals, and 223 reptiles (Petocz & de Fretes, 1983; Petocz, Kirenius, & de Fretes, 1983; Petocz, 1984, 1989; BAPPENAS, 1991; WWF, 1993a; Petocz & Raspado, 1994). Indonesia

Fig. 7.1 The island of New Guinea

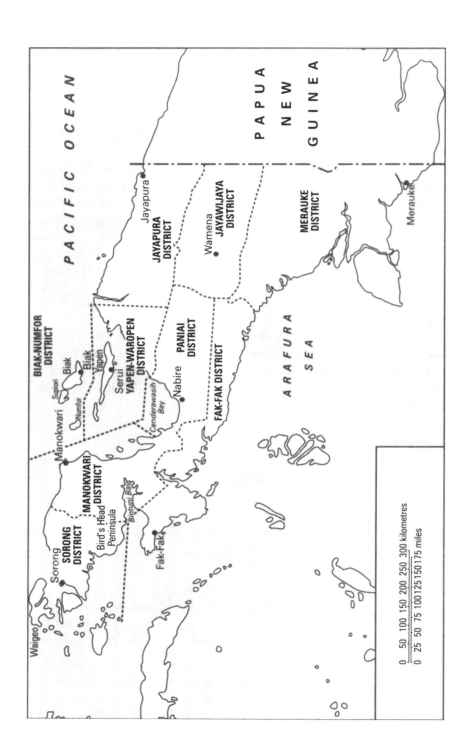

PACIFIC OCEAN

PAPUA NEW GUINEA

Jayapura

JAYAPURA
DISTRICT

Wamena
JAYAWIJAYA
DISTRICT

MERAUKE
DISTRICT

Merauke

BIAK-NUMFOR
DISTRICT

Biak
Biak
Yapen

Supiori

Serui
YAPEN-WAROPEN
DISTRICT

Numfor

Nabire
PANIAI
DISTRICT

Cenderawasih
Bay

ARAFURA
SEA

FAK-FAK DISTRICT

Manokwari

MANOKWARI DISTRICT

Bird's Head
Peninsula

Bintuni Bay

Sorong
SORONG
DISTRICT

Fak-Fak

Waigeo

0   50 100 150 200 250 300 kilometres
0 25 50 75 100 125 150 175 miles

has up to about 15% of all the terrestrial species in the world (Chapter 4), and Irian Jaya may well contain half of them. The seas around New Guinea are also extremely species-rich, and perhaps more so that anywhere else (Norse, 1993; Huston, 1994). In terms of biological diversity, therefore, Irian Jaya can reasonably be described as the single most important part of one of the most important countries in the world.

7.1.2 Background to conservation action

Conservation planning in Irian Jaya during the late 1970s and early 1980s resulted in the creation of a province-wide system of nature reserves (MacKinnon & Artha, 1981; Petocz, 1983; Diamond, 1986). These were chosen to ensure the inclusion of rare and representative local ecosystems, and some 55 reserves had been constituted or formally proposed by the late 1980s (Nash, 1991). These reserves had been incorporated in provincial spatial planning maps, so that planned development projects have generally been located outside their boundaries. As areas outside the reserves were allocated to other uses, however, there was increasingly less scope to amend reserve boundaries in response to new information on the distribution of species and ecosystems. The existing reserves were seen as the locations where natural ecosystems were most likely to be saved intact, and all other areas were assumed to be at serious risk of disturbance or conversion.

During most of the 1980s, and continuously since the late 1980s, WWF has been involved in Irian Jaya as the principal adviser on conservation issues to the Ministry of Forestry through PHPA and its technical arm KSDA (see Chapter 4 and the Glossary), as well as to other branches of national and local government. By 1994, WWF had accumulated in Irian Jaya many experienced staff and resources such as library materials, buildings, road vehicles, boats, and other equipment, partly as a result of long-term joint-funding by the British Government since 1990. These resources were being used to support a range of site-specific field projects and province-wide education activities, as well as to provide advisory services in the development of government policies.

Projects undertaken by WWF in Irian Jaya have concentrated on parts of the reserve system, and have emphasized collaboration with the people living

Fig. 7.2 Irian Jaya

in and around those reserves. These projects were designed to recognize a common theme in preserving biological and cultural diversity, which is that people and wild species have a similar need for secure access to the resources with which to sustain their culture and life. This means stabilizing the boundaries of reserves, both to allow viable natural ecosystems to survive, and also to make it possible to work with viable local communities. The latter require clear rights to occupy territory and to use resources in and around reserves, if they are to be involved in the management of those reserves.

These projects aimed to help people living around reserves to participate in setting their boundaries, in harvesting certain natural resources within them, and in gaining benefit from new forms of sustainable economic activity there (Nash, 1991; Worah, 1994). It was learned from projects in Irian Jaya and elsewhere during the early 1990s that mapping of individual and community land holdings around reserves can help to stabilize their boundaries (WWF, 1991a, 1991b, 1993a, 1993b, 1993c, 1993d, 1994a, 1994b, 1994c, 1994d; Stark, 1992a, 1992b). This was attributed to the effect of clarifying resource tenure in helping people to avoid competitive or open-access exploitation of resources, and in encouraging investment in long-term resource management.

It had proved feasible to work with local people in agreeing boundaries, in marking them on the ground, and in mapping the agreed boundaries for submission to the Ministry of Forestry. By late 1994, the ministry was becoming willing to consider accepting locally determined boundaries in preference to the original, centrally planned ones. Events were thus moving into a new phase in Irian Jaya, and a review of progress had become needed to allow the formulation of a plan of work for the period 1995–2005 (Caldecott, 1994a; WWF, 1995a).

Such a review needed to be guided by the quickly changing circumstances of Irian Jaya's system of nature reserves. A study of these in late 1994 provided information on the kinds of threat to biological and cultural diversity which were becoming dominant at that time in Irian Jaya (see below). These threats ranged from road building, logging, fishing, and mining, to the development of plantation agriculture and the introduction of large numbers of non-aboriginal people through official transmigration schemes and spontaneous migration. By 1994, the number of transmigrants in Irian Jaya had exceeded the number of aboriginal people living there, as predicted by Diamond (1986).

An overall impression gained by the review in late 1994 was that Irian Jaya was at the beginning of a process of economic development which would transform the province over the next 10–20 years. The likely course of this development is shown by what had happened in the previous three decades in western and central Indonesia, and elsewhere in South-east Asia (Chapters 2, 4, & 5). The general pattern had been for forests to be degraded and destroyed, for much native biodiversity to be eroded and lost, and for many distinctive local cultures to be erased. In Irian Jaya in the mid-1990s, conservationists still had the opportunity to imagine a different kind of future, but they also had the challenge of how to achieve any part of it in the prevailing political and economic circumstances.

7.2 Threats and action in the field

7.2.1 Cyclops Mountains

The forests of the Cyclops or Dafonsoro Mountains are located close to Irian Jaya's capital city of Jayapura (Fig. 7.3). They were originally protected within a 225 km^2 Strict Nature Reserve (SNR) for three main reasons: that they comprise a water catchment for Jayapura, that they grow on an unusual ultrabasic rock formation and shelter several endemic species, and that their proximity to Jayapura gives them potential value in research, recreation, and education. These forests were threatened in late 1994 by the encroachment of small-scale farmers from the surrounding lowlands. This process was being driven by increasing settlement of newly arrived people in the area, and by uncertainty over land ownership and the precise location of reserve boundaries.

Work had been undertaken by WWF since 1986 with aboriginal communities around the reserve to agree, mark, and map reserve boundaries. Some new techniques of land management (e.g. pond aquaculture and agroforestry) had been introduced in an effort to increase the efficiency of local use of land, and thus to relieve land shortage. After some initial reluctance, the new boundaries had been accepted by the Forest Department, and in general appeared to be stable. A building had been rehabilitated jointly with the community at Pos Tujuh, near the boundary of the reserve, to be used as a field base for conservation staff of both WWF and PHPA/KSDA, and as a visitor interpretation centre.

Coral reefs along the shore adjacent to the northern boundary of the reserve had been damaged by the use of explosives for fishing. This is a common and widespread problem in eastern Indonesia and Irian Jaya which is rapidly destroying the region's marine and coastal resources. It is often combined with the use of cyanide for fishing and the dredging of reefs for building and as a source of lime. Since 1990 in the Cyclops area, WWF had responded

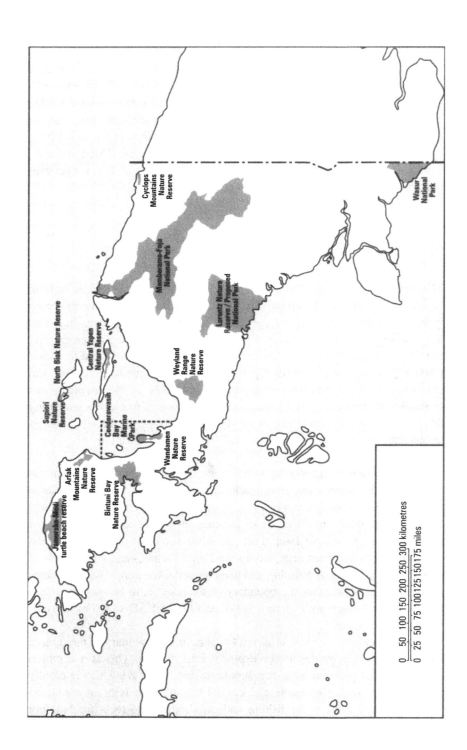

Cyclops
Mountains
Nature
Reserve

Wasur
National
Park

Mamberamo-Foja
National Park

Lorentz Nature
Reserve / Proposed
National Park

North Biak Nature Reserve

Central Yapen
Nature Reserve

Weyland
Range
Nature
Reserve

Sapiori
Nature
Reserve

Cenderawasih
Bay:
Marine
(Park)

Wandamen
Nature
Reserve

Arfak
Mountains
Nature
Reserve

Jamursba-Medi
turtle beach reserve

Bintuni Bay
Nature Reserve

```
0   50  100 150 200 250 300 kilometres
|___|___|___|___|___|___|___|
0   25  50  75 100 125 150 175 miles
```

by promoting local awareness of the unsustainable and illegal nature of fish bombing, and had provided a boat to help local people protect their resources. Communities had discussed the issues involved, and had developed local rules and sanctions for regulating their own behaviour. These efforts had been supplemented by government patrols to combat bombing by outsiders, and the incidence of bombing had been greatly reduced by 1994.

7.2.2 Biak and Supiori

The island of Biak, to which Supiori is linked by a narrow isthmus, lies in the Pacific Ocean and has a coralline limestone geology reflecting its distance and isolation from New Guinea. This isolation has resulted in Biak-Supiori having the highest level of endemism per unit area in Irian Jaya, for example with nine endemic bird species and many endemic subspecies (ICBP, 1992, 1995). The poor soils mean that farming is unproductive and that there is little regeneration after logging. Nevertheless, most of Biak was logged during the 1970s, and farmed during the 1980s, and by 1994 was dominated by derived heath forest and scrub, which seemed likely to persist indefinitely. The forests in part of north Biak and in most of Supiori had survived into late 1994, however, due to the rugged terrain on which they grew, and they were contained within the North Biak and Supiori SNRs.

Although roads were built around the northern boundaries of both reserves in the late 1980s, it was not until 1994 that extensive forest clearance began along these roads. This resulted from the fear of local people that new settlers were about to occupy land which they perceived as their own. Local residents were therefore attempting to establish secure tenure by clearing and planting areas adjacent to the reserves. This was damaging the reserves themselves, because of uncertainty over the location of their boundaries, and also threatened wild species that depended on access to lowland and coastal habitats outside the reserves.

The Supiori reserve was also threatened by construction of about 20 km of additional roads, around its eastern and southern boundaries. These new roads also seemed likely to encourage land clearance by aboriginal people in anticipation of conflict with new settlers. A plan had been devised by WWF to work with local communities around both reserves to clarify land

Fig. 7.3 Conservation areas of Irian Jaya

ownership and to map and mark reserve boundaries. The aim was to re-establish local confidence in land tenure, in the hope that residents would become willing to adopt more benign ways of using land than the traditional practice of burning forest and planting corn and yams.

7.2.3 Yapen

The island of Yapen lies to the south of Biak and is much closer to the New Guinea coast, to which it was connected in the recent geological past. The island therefore has few endemic taxa, but in late 1994 it was still almost entirely forested and was very rich in species (with at least 134 birds, 28 butterflies, 69 orchids, and 32 mammals). Yapen therefore represented a large and important sample of the natural ecosystems of northern Irian Jaya. By 1994, however, logging was well advanced in the adjacent Mamberamo region of the mainland (McCarthy, 1991), and the possibility had been raised officially of damming the Mamberamo river for hydro-electricity. This implied that Yapen was likely to become an increasingly isolated refuge of intact, natural vegetation. With this role as a sample in mind, and in view of the steep terrain, a large part of the interior of the island had been included in the 590 km² Central Yapen SNR, and most of the rest was classified as Protection Forest by RePPProT and TGHK. These classifications were retained in the provincial spatial plans of 1991 and 1992.

The community of Ambaidiru had existed within the SNR since before the reserve was created, and WWF and PHPA/KSDA had proposed to work with the inhabitants to define the boundaries of a suitable enclave. A greater threat by 1994 was posed by several roads which had recently been completed, one between the north and south coasts of Yapen directly through the reserve, another inland from the south coast, and others were planned as far as the eastern extremity of Yapen. Farmers were already clearing land within the reserve along these new roads. The roads were being financed by the Provincial Government, but without complying with the environmental impact assessment and management (AMDAL) procedures specified by Government Regulation 51 of 1993 (BAPEDAL, 1993, 1994). The District Government was aware of the need for AMDAL where proposed roads enter an SNR, but no funds had been made available with which to undertake the process. In any case, there was clearly a strong commitment by local government to build roads on Yapen, since this was seen by officials as the best way to develop the island quickly.

Events like those on Yapen were by 1994 becoming common in Irian Jaya, as large sums for road building (but only rarely for AMDAL) were being spent by government, and as confusion persisted over the applicability of the AMDAL process (C. Vandersluys, personal communication, September 1994). In the case of Yapen, WWF had been active there since 1992, and had proposed to continue working with local people to clarify, mark, and map agreed external and internal boundaries of the reserve. The new road system threatened to overwhelm all such efforts, and had already caused widespread damage. One option that was discussed informally with local officials was to make the whole island a National Park, which would allow the island to be zoned more flexibly than before. This change would perhaps make it easier for PHPA/KSDA to protect large areas of the interior and coastline, and to suppress commercial hunting of protected species such as parrots and birds of paradise. Compromises of this sort were by 1994 becoming typical of conservation work in Irian Jaya.

7.2.4 Arfak Mountains

The coastal Arfak Mountains lie a few kilometres south of the town of Manokwari and rise to nearly 3000 m. The 683 km^2 Arfak SNR contained generally intact, natural forest in which lived at least 70 species of mammals and 320 birds. Arfak is part of an endemic bird area (ICBP, 1992, 1995), and a centre of diversity for birdwing butterflies (including the local endemic *Ornithoptera rothschildi*), which feature prominently in WWF's management strategy for the reserve (Craven & de Fretes, 1987). In late 1994, threats to the reserve were mainly side-effects of the poverty of aboriginal people in the area, or else involved illegal hunting and gathering by outsiders, and small-scale encroachment across uncertain reserve boundaries. New roads were also planned which would approach the reserve in the north and south-west.

Conservation work in the area since 1987 had focussed on working with local Hatam people to mark and map agreed boundaries of the reserve (Craven, 1989). This process had yielded boundaries which were acknowledged by the Ministry of Forestry's Bureau of Forest Inventory and Mapping (BIPHUT) to be more stable and durable than the former, centrally planned boundaries. Co-operative shops and saving schemes had also been established in local villages, with the aim of teaching management skills and promoting local self-reliance and financial security. Terraced farming techniques had been promoted by local WWF employees, and a project had been developed in which WWF was working with a local NGO to promote the ranching, harvesting, and sale overseas of protected birdwing butterflies (Neville, 1992; New, 1994).

These activities appeared to have been effective in helping to make the Arfak Mountains reserve 'real', both on the ground and in the minds of the local people. There were several problems, although these were largely beyond WWF's control. They included the threat of road construction, and also the vulnerability of the butterfly trading operation to the certification requirements of the Convention on International Trade in Endangered Species of Wild Fauna and Flora (CITES). These were time-consuming to fulfil according to Indonesian procedures, and there was competition from other traders who were evading CITES requirements illegally. A way round the latter problem was being sought by 'ecolabelling' butterflies for domestic sale, and by encouraging foreign customs departments to accept those labels as satisfactory proof of CITES compliance.

7.2.5 Cenderawasih Bay

Cenderawasih Bay lies on the northern coast of Irian Jaya, south of Biak and Yapen and between the Bird's Head Peninsula and the main body of New Guinea (Fig. 7.3). The 14 535 km² Cenderawasih Bay Marine Park was established in the western side of the Bay, between the coast and arbitrary boundaries out to sea. The larger, northern section of the Park was in Manokwari District and contained four inhabited islands, while the rest was in Paniai District and included seven coastal villages. The total population in 1994 comprised about 12 600 people. A chain of small, uninhabited islands and reefs stretched north-south for the length of the Park. A large peninsula at the southern end of the Park was mostly included within the Wandamen SNR.

Preliminary surveys in Cenderawasih Bay had identified 241 species of coral, 355 fish, and 196 molluscs, but far more were expected to be present given the area's location in the species-rich waters of the central Indo-Pacific. Threats to the marine and coastal resources of the Park in 1994 were similar to those elsewhere in the waters off Irian Jaya. They included fishing using explosives and potassium cyanide, collection and sale of protected species (e.g. giant clams, ornamental shells, turtles and dugongs), and incursion by fishing boats from outside the area. The latter were coming from as far away as Java and Sulawesi, but were likely soon to be supplemented by boats supporting a new fish cannery in Manokwari town. These activities threatened to damage marine ecosystems and wildlife populations, making them unable to support productive harvests or attract tourists. Tourism was an important management aim of the Park, but the devastated reefs around Biak and

Yapen, and in some places in the Park itself, would take several years to recover even if fully protected. An indirect threat to the Park was logging inland to the south, which increased sedimentation in the marine environment and meant that rafts of logs traversed the Park on the way to Biak.

A project had been developed since 1992 in the southern, Paniai sector of the Park, involving WWF and PHPA/KSDA in working with groups of fishers in the seven villages in the area. The aim was to increase awareness of resource protection regulations, while also promoting co-operative ventures based on fish processing, livestock, and fruit trees. This assistance was intended to help local people adjust to the loss of opportunities to sell protected marine organisms. Their motivation to forego using these resources, however, was being affected by seeing the same resources being taken by outsiders. There was thus a need both for more active patrolling of the waters within the Park, and also for clear and exclusive rights for residents of the Park to use resources there.

The need for patrols in the Paniai sector was met in September 1994 by means of a task force created by the Paniai District Government. This included members of the armed forces and police as well as PHPA/KSDA, and made use of a boat provided by WWF. The right of local people to harvest marine produce within the Park in accordance with management regulations was already accepted in practice by local government, although would need to be confirmed by the Ministry of Forestry. This combination of measures was expected to improve protection of the Paniai sector of the Park, and provide a model for later extension to the larger Manokwari sector.

7.2.6 Bintuni Bay

Bintuni Bay lies on the western side of the isthmus which joins the Bird's Head Peninsula to the mainland of New Guinea, on the far side of the watershed from Cenderawasih Bay (McCarthy, 1991). In late 1994, it still contained one of the largest areas of mangrove in the world, at about 5000 km^2 similar in extent to those of the Niger Delta in Nigeria, and about half of these mangroves were contained within the Bintuni Bay SNR. According to the Manokwari District Development Planning Agency (BAPPEDA), however, the Bay was vulnerable to use by groups with interests in exploiting many different kinds of resources. These included aboriginal fishing communities as well as transmigrant fishers from Java, and commercial concerns focussing variously on fish and prawns, sago palms, wood pulp, oil, and natural gas.

Considerable damage had been done to the ecosystems of the area, with impacts on local aboriginal people (SKEPHI, 1990). One positive development, however, was that the boundaries of an SNR in part of the area had been clarified relative to an adjacent wood pulp concession, and the company concerned had been fined and compelled to replant part of the reserve which had been damaged by its operations. Another was that the multiple competing fishing companies which formerly operated in the Bay out of Sorong had been replaced by a single company, with authority from the Ministries of Forestry and Fisheries. Nevertheless, conflicts of interest abounded in the area, and in September 1994 the Manokwari BAPPEDA indicated that it would welcome work by WWF in support of their own detailed spatial plan for the Bay, and in the context of an existing integrated resource plan for the District as a whole (Hunt, 1992).

7.2.7 Jamursba-Medi

In the early 1980s, there were five major nesting beaches of the endangered leatherback turtle (*Dermochelys coriacea*) on the northern coast of the Bird's Head Peninsula. By the early 1990s, however, killing of turtles and predation on eggs by people and exotic pigs (*Sus scrofa*) had reduced this number to one, Jamursba-Medi, with only isolated cases of nesting elsewhere. Monitoring of nesting by PHPA/KSDA staff in 1990–4 put the number of leatherbacks using Jamursba-Medi at 1000–1600 each year. This implied that use of the beach was a significant factor in the survival of the leatherback population of the western Pacific.

In addition to direct predation on the turtles and their eggs, other threats in 1994 derived from the fact that the whole area and its hinterland was classified as production forest, and was being exploited for timber. Cases were known of leatherbacks being slaughtered by timber workers using chainsaws, while roads and log-loading activity generated other dangers. A gold-mining company had also surveyed the area, and the landing of heavy equipment or establishment of facilities on or near the beach was considered an imminent threat in late 1994, and later became a real one (R. A. Betts, personal communication, March 1995).

The response by WWF and PHPA/KSDA was to employ local people as turtle guards, while also promoting awareness of turtle protection among the inhabitants of two local villages. A rubber boat had been provided to support patrols against the harvesting of eggs by outsiders. Local means of subsistence were being promoted by WWF as an alternative to predating turtles, especially by encouraging the sale of palm sugar to coastal traders, and by demonstrating vegetable gardens. Direct turtle protection measures included relocating nests from vulnerable locations, protecting nests within pig-proof cages, and head-starting hatchlings on a trial basis. A proposal had also been made to upgrade the status of the beach to that of Wildlife Reserve, with an area of about 100 km².

7.2.8 Wasur

The Wasur National Park lies between the Maro River and the PNG frontier in the extreme south-east corner of Irian Jaya, close to the town of Merauke to the west and bounded by the Arafura Sea to the south. The climate is strongly seasonal, with a single rainy season, and the area is dominated by savannah woodlands, with patchy stands of moister *Eucalyptus* and *Melaleuca* forest, and large areas of seasonal swamp and grassland. Native wildlife species include at least 390 birds, and the area is part of an endemic bird area (ICBP, 1992, 1995) as well as being used by many migratory waterbirds such as ibises and pelicans. There are some 80 native mammals, including abundant wallabies (*Dorcopsis veterum* and *Macropus agilis*), and exotic deer (*Cervus timorensis*) and pigs are also common. These habitat associations and wildlife resources occur nowhere else in Irian Jaya or elsewhere in Indonesia, and by 1994 had attracted government interest and investment, especially in their tourism potential.

Several threats to the Park and nearby ecosystems had become established by late 1994. Firstly, transmigration schemes opened successively during the 1960s, 1970s, and 1980s had isolated the Park behind a barrier of settlements and rice fields. Secondly, there was intensive hunting to satisfy the demand for wild meat created by the influx of transmigrants, and by the growth of Merauke town. Thirdly, aboriginal people had been displaced from areas far from the Park into its buffer zone, which was a narrow band of forest mostly between the transmigration schemes and the Maro River. Lastly, the Trans-Irian Highway had been constructed through the middle of the Park and then northwards along the PNG frontier, and several thousand non-aboriginal people had been settled along it and around Sota on the PNG frontier deep in the Park.

The Park itself occupied the traditional lands of three aboriginal groups, who were based in a number of villages inside the Park. These communities had always been seen by WWF as allies in the management of the Park, rather than for example as candidates for resettlement (I. Craven, personal communication, February 1992). This had strongly influenced WWF's approach in designing and implementing the Wasur project. Its first aim was to establish the right of aboriginal people to live within the Park and to harvest its resources in accordance with the management plan (PHPA, 1992). The project then sought to work with PHPA/KSDA to prevent outsiders hunting within the Park, while promoting the sale of deer and pig meat to outsiders by traditional hunters operating within the Park. It also sought to promote village-level production and wider sale of aromatic oil distilled from *Eucalyptus* leaves, and to encourage community tourism initiatives.

The Wasur project had worked with communities in the buffer zone of the Park, in order to clarify tenure over resources, and to stabilize forests as far as possible within traditional use zones owned and managed by aboriginal groups displaced by transmigration schemes. It had encouraged government to resettle non-aboriginal communities along the Trans-Irian Highway to less sensitive locations, and to close the Highway itself to public traffic while upgrading the alternative road through transmigration sites outside the Park. These aims had mostly been achieved by late 1994, largely due to the efforts of the late Ian Craven, and WWF's involvement in Wasur had therefore entered its final phase. The main theme of this was to ensure a smooth transition of the Park to full PHPA/KSDA management, which was due to be completed in 1996, while also keeping the Park's aboriginal residents involved in its management. The procedures and precedents established during this process are likely to be applicable in many other locations in Irian Jaya and elsewhere in Indonesia.

7.2.9 Lorentz

In late 1994, the Lorentz SNR remained one of the most important terrestrial nature reserves in the equatorial tropics, containing a vast sample of Irian Jaya's natural ecosystems, most of which remained intact. It extended from the mangroves of the southern coastline of Irian Jaya on the Arafura Sea, inland through swamps and lowland forests, and then steeply upward to the crest of the Central Cordillera, where it embraced glaciers, alpine meadows, and the peak of Mt Carstensz (Mt Jaya) at 4884 m. All of Irian Jaya's major ecosystem types and many of its vegetation sub-types were included in the reserve, which sheltered about two-thirds of Irian Jaya's known bird species (20 of them endemic to Lorentz) and four-fifths of its mammals (WWF, 1993a). The Lorentz area is listed as a centre of plant diversity (WWF & IUCN, 1995) and includes parts of two endemic bird areas (ICBP, 1992, 1995). Lorentz has been inhabited for at least 30 000 years, and in the early 1990s it contained about 11 000 residents belonging to eight tribal peoples, the Nduga, West Dani, Sempan, Amungme, Komoro, Ekagi, Asmat, and

Nakai (Manembu, 1991). It is thus hard to overstate the role of Lorentz in supporting cultural and biological diversity.

The Lorentz area was originally declared a nature reserve in 1916, and it was gazetted by the Indonesian Government in 1978 as an SNR with an area of about 21 500 km². It was later proposed as a National Park with an area of about 15 000 km², the size being reduced to exclude settled areas in the highlands as well as an area affected by mining interests (Petocz, 1983; WWF, 1993a). Although there was broad official acceptance of the idea of National Park status for Lorentz, its status, size, and boundaries were still uncertain in late 1994. By that time, WWF had been active in the area for several years, mainly in documenting local cultures and laying the groundwork for later management by building links with local people and assessing access routes and potential locations for research and infrastructure sites. Most of the eastern and western boundaries of the reserve had meanwhile been surveyed and marked by BIPHUT.

Threats to Lorentz in 1994 included illegal logging which had occurred in the south-eastern part of the reserve, oil exploration, and medium-term government plans to drive roads through the area. The greatest source of uncertainty over the reserve's status, however, was the presence of an enormous copper and gold mine located high in the mountains near the north-western boundary of the reserve (PaVo, undated a & b). The company concerned, the Indonesian subsidiary of a large trans-national corporation, Freeport-McMoRan, had exploration rights to large areas of the reserve, and significant ore deposits had been discovered there. Experience outside the reserve suggested that open-cast mining in the upper watershed of the reserve itself would cause damage through downstream siltation, unless mining operations were redesigned. There was also the likelihood of friction with aboriginal people, and possible violence between the Indonesian armed forces and the Independent Papua Organization (OPM).

The status of the Lorentz reserve in the mid-1990s thus had the potential to become an international *cause celèbre* on the scale of logging in Sarawak in the late 1980s. In late 1994, real progress seemed likely to depend on achieving a viable political consensus within the Indonesian Government that the reserve should be made a National Park and therefore not available for mineral exploitation. This consensus would presumably depend on the mining company accepting the loss of access to some ore bodies, combined with the use of new technology to prevent river-borne sediment entering the new park, and perhaps some adjustment of its boundaries. Thereafter, management of the park would depend on close collaboration with its aboriginal inhabitants, including the mapping of community land holdings, and the development of protection, research and tourism capacities by PHPA. Because of the nature of the underlying problems, the safety of Lorentz seemed unlikely to be achieved without reaching an agreement with groups outside the project area. In 1995, WWF and the German Government agency GTZ began a process of discussion, consensus building, and field work aimed at achieving this outcome.

7.3. The project design

7.3.1 Lessons from the field examples

The accounts in section 7.2 illustrate many of the threats to biological and cultural diversity which were active in Irian Jaya in 1994. Infrastructure development was occurring without adequate prior assessment of environmental and social risks or consideration of alternatives (e.g. in Yapen, Arfak, and Biak-Supiori). There were cases of logging and mineral concessions which overlapped, abutted, or impacted on nature reserves (e.g. in Bintuni, Cenderawasih, Jamursba-Medi, and Lorentz). There were signs of destructive and unsustainable uses of wildlife resources by non-resident outsiders, both marine (e.g. in Biak, Yapen, and Cenderawasih) and terrestrial (e.g. in Cyclops, Arfak, Biak-Supiori, and Wasur). Farming encroachment was common due to uncertainty over land tenure and the precise boundaries of reserves (e.g. in Cyclops, Arfak, Biak-Supiori, and Yapen). Lastly, there were the direct and indirect effects of transmigration schemes (e.g. in Wasur and Bintuni).

Results from WWF's projects in Irian Jaya suggested that useful progress could be made in several ways. Firstly, it was possible to work with aboriginal and other local people to agree, mark and map the boundaries of nature reserves, buffer zones and land-holdings within those buffer zones. Equivalent improvements in tenure over marine and coastal resources could also be made. Secondly, the same people could be involved in harvesting and marketing local products, and in saving and productively using revenues obtained in these ways. Lastly, it had also proved effective to work closely with PHPA/ KSDA, both to support its own growing capacity to protect species and ecosystems in the field, and to enhance links with other agencies of government and local people. Some lessons learned from projects in Irian Jaya have been consolidated into a set of guidelines for involving local people in the management of nature reserves (WWF, 1993a), and these are summarized in Box 7.1.

Box 7.1 Guidelines for involving local people in reserve management (adapted from WWF, 1993a).

Guideline 1: The development of management activities must be sustainable and should contribute directly and locally to conservation.

Guideline 2: Activities should either significantly increase incomes or reduce demands on labour inputs.

Guideline 3: Benefits should be clearly achievable in the short term, and this should be apparent to people involved in the project.

Guideline 4: Activities should be clearly linked to the protected area either directly (e.g. a critical resource found there is conserved) or indirectly (e.g. development in the buffer zone or use of park resources is contractually dependent on actions taken to protect the park).

Guideline 5: Actions should be diversified in order to minimize competition within communities and to avoid overproduction for local markets.

Guideline 6: Markets for products should be nearby, so as to minimize transport costs and difficulties of distribution.

Guideline 7: Benefits should be spread evenly to minimize jealousies within and between communities, and it is often better to divide available resources among various projects and communities rather than use them all in one large demonstration.

Guideline 8: The scale of management activities should be large enough to provide tangible benefits within communities, but not so large as to attract unwanted outside entrepreneurs.

Guideline 9: Community organizations should not be overloaded with too many projects at any given time and place, and where a range of compatible activities occur in a region these should not have to compete with one another for resources.

Guideline 10: There should be no direct grants of cash or equipment without a clear expectation of return from the recipients, such as maintenance of reserve boundaries and/or resources, and recipients should never be paid wages to perform any activity that directly and clearly benefits them without such payment.

Some threats were originating with planning and investment decisions over which the communities involved in conservation projects had little or no influence. The AMDAL process was designed to provide some defence against initiatives which use or impact upon resources in ways other than planned. Although it cannot guarantee the security of nature reserves, it can prevent them being destroyed through carelessness, while also providing an opportunity for public opinion and technical advice to exert influence on the decisions involved.

This defence could not be used effectively in Irian Jaya, however, because the AMDAL process was not being used routinely prior to project approval and implementation. There were several reasons for this, including a lack of awareness of the law and a lack of resources with which to undertake AMDAL studies, and these add up to a major set of constraints on effective conservation in Irian Jaya. Making it possible for developers to apply the AMDAL process routinely was therefore seen as an important next step in the WWF Irian Jaya programme. This was intended to complement both WWF's proven tenure-based and community-based approach to project design, and also the government's existing spatial planning framework for the Province.

7.3.2 *Legislative and policy framework*

Policies and laws cannot on their own ensure sustainable use of resources, but the trend in Indonesian government thinking in 1990 to 1994 increasingly supported conservation. Act 5 of 1990 defined the nature and management objectives of the various kinds of nature reserve, while Act 24 of 1992 created a legal framework for national and regional spatial planning. Presidential Decree 55 of 1993 then recognized communally owned lands and provided for compensation in respect of their acquisition. Government Regulation 51 of 1993 clarified procedures for analyzing and managing the environmental and social impacts of development projects (collectively referred to as the AMDAL process).

All projects which might impact upon a protected area must undergo AMDAL before a decision is made to proceed with them. This applies to the various kinds of Nature Conservation Areas (National Parks, Grand Forest Parks, and Natural Recreation Parks) and Sanctuary Reserves (Strict Nature Reserves and Wildlife Sanctuaries) defined under Act 5 of 1990. It also applies to Protection Forests and a wide range of sensitive terrestrial and aquatic ecosystems. These include peat and mangrove areas, water catchments, coastal and riverine edges, areas surrounding lakes, reservoirs, and springs, aquatic ecosystems with high biodiversity, and other areas with special attributes.

In June 1994, the Minister of Forestry made a statement which stressed the importance of biodiversity and effectively authorized local and traditional community involvement in sustainable forest management. Thus, by late 1994, the Ministry of Forestry was in a position to accept as valid the maps of reserve boundaries prepared by local communities with WWF's assistance. There was also an unambiguous legal requirement to complete the AMDAL process before implementing any development project likely to affect a nature reserve. These improvements were reflected in the design of the next phase of WWF's involvement in Irian Jaya, along with the conclusions of a number of strategic analyses which WWF had completed over the same period (WWF, 1991c, 1993a, 1993d, 1994c, 1994d).

7.3.3 Overview of the new programme

The overall aims of WWF in Irian Jaya are reflected in the kinds of field activity described above. They centre on helping the Ministry of Forestry to manage reserves, promoting awareness and support for conservation, collecting information about ecosystems and wild species, and developing human resources for conservation through field experience, training, and research. The 1994 proposals for work by WWF in Irian Jaya in the period 1995–2005 were described as 'more of the same, but better, plus AMDAL' (R. A. Betts, personal communication, September 1994). The addition of AMDAL as a major theme was made in response to increasing competition for land and other resources between those wishing to conserve biological and cultural diversity, and all other groups. Pressures to exploit natural ecosystems were expected to increase rapidly, with drastic effects on the people dependent on them. Irian Jaya's reserve system would become vulnerable to both planned and unplanned damage. Under the kinds of threat illustrated above, it was no longer considered sufficient to declare the intention to conserve, or to present inaccurate maps of nature reserves as if those areas had been conserved. More specific action was urgently needed, which was to be undertaken through a new programme designed around four components (Fig. 7.4).

Field teams would work with PHPA/KSDA and local people around priority reserves to clarify ownership of resources, to mark boundaries physically and to map them accurately using global positioning system (GPS) and geographical information system (GIS) units (see Glossary). The field teams would also develop proposals for more elaborate projects where necessary, and feed information back to Jayapura. Meanwhile, information managers based in Jayapura would compile and use field, literature, and mapped data to anticipate conflicts with conservation priorities, and to make it easier for proponents of development activities to use the AMDAL process to avoid those conflicts. Information distributors would also work at schools and transmigration sites throughout Irian Jaya, and at the Pos Tujuh field centre near Jayapura, to promote understanding of conservation issues, and to reinforce public support for the protected area system. Lastly, programme managers

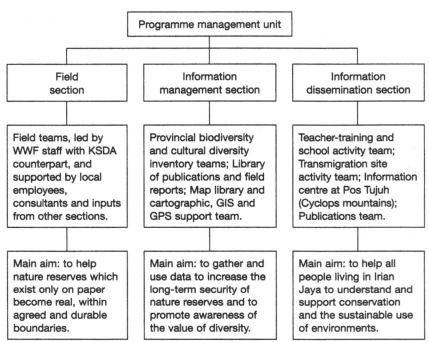

Figure 7.4: Structure of the proposed WWF Irian Jaya programme.
Source: Caldecott, 1994a; WWF, 1995a.

based in Jayapura would administer and support all activities while explaining and representing the programme to government.

7.4 Discussion

The circumstances of Irian Jaya in the first half of the 1990s were reminiscent of those in the interior of Sarawak 10 years before (Chapter 2). Both territories were laden with valuable natural resources which could be exploited profitably using technology and capital resources which were already available. Both were sparsely populated by aboriginal people with little direct influence over the decisions of central government and big business. In both cases, final allocations of forest lands were being made to different kinds of use, for protection, production, or conversion. It is possible, however, that events will unfold somewhat differently in Irian Jaya than in Sarawak, although there is no foreseeable outcome which will not involve significant loss of wildernesses, cultures, languages, and species.

The few years following 1995 were expected to be crucial to the future of the environment in Irian Jaya, and conservation groups such as WWF will help to determine what happens within broad limits. Despite overwhelming threats, influence can be exerted on behalf of diversity by using and building upon WWF's existing assets. These include abundant, motivated and experienced staff, and the intimate relations with government and local people which have been established throughout the Province over the decade between the mid-1980s and mid-1990s.

These resources have been used to help create a policy dynamic within the Indonesian Government, leading to attitudes and laws which are increasingly favourable to preserving at least the main nature reserves of Irian Jaya. These reserves were selected specifically to ensure that most native species would survive within them, provided threats could be resisted successfully. One way to do so is to ensure that their boundaries are marked and mapped, and that they are actively protected by the people who live around them. The value of this approach against certain kinds of threat has been shown by WWF's recent field programmes.

Another way to do so is to ensure that the provisions of Indonesian law are applied so as to avoid the accidental destruction of protected areas. This approach is needed to guard against threats which cannot be neutralized through community action alone. Both approaches were feasible by 1994 in terms of the motivation of local people, the direction of law and government policy, the availability of appropriate technology, and the skills of conservationists. What is certain, however, is that conservation in Irian Jaya will be hugely expensive, and that the necessary investment is needed in the period 1995–2005, rather than at some point in the future when there will be far less left to save.

8

Project themes and practicalities

8.1 Introduction

Chapters 2–7 summarized field experience in designing conservation projects in Sarawak (Malaysian Borneo), Cross River (Nigeria), Siberut and Flores (western and central Indonesia), Hainan (China), Luzon (the Philippines), Costa Rica (Central America), and Irian Jaya (Indonesian New Guinea). These are described in more or less the same order in which the projects occurred in 1984–94. This chapter aims to review the underlying conservation problems in each place, and the nature of the solutions attempted there. It then draws on practical lessons learned in designing the various projects, in order to frame some general guidelines which may be useful to other conservationists faced with similar sets of problems.

Threats to biodiversity can often be traced back to one or more underlying problems. These may be linked to the failure of planning, markets, policies, or institutions, to distorted distributions of wealth and power or poverty and weakness, to excessive numbers of people, or to open-access exploitation of renewable resources. These conclusions were reached partly as a result of the work described in Chapters 2–7, and examples include:

- planning failure in Siberut and Biak, where forests had been allocated to logging despite the steepness of the terrain, wet climate and/or extremely fragile soils;
- policy failure in Makiling Forest Reserve, where confused management aims had encouraged colonization and forest clearance;
- market failure in Sarawak and Siberut, where the external costs of logging had been borne not by logging companies but by forest-dependent people;
- institutional failure in Yapen, Irian Jaya, where roads were being built through a Strict Nature Reserve without compliance with EIA procedures; and

- open-access exploitation of wild meat and plants in Cross River National Park, Makiling Forest Reserve, and Bawangling Nature Reserve, and of fish in Cenderawasih Bay.

Serious poverty was a rare factor in the cases described, since all were set in lower-to-middle income countries rather than in poor ones (UNDP, 1991). Relative poverty was more of an issue, since the projects were set in rural areas which were in most cases far from economic centres. This was often responsible for the existence of each project, since relative poverty had drawn the interest of development agencies, while surviving natural ecosystems had attracted conservation ones. The places described were thus distinctive parts of each country, and these special conditions became more marked at the level of districts, small towns, and villages located near reserves, which tended to be isolated or marginal to the national economy.

8.2 Overview of the case studies

8.2.1 Sarawak, 1984–1988

Sarawak had a pioneer economy, with few roads and much remaining forest, a low overall population density, and many traditional, forest-dependent people, as well as some who were forest-dwelling. Its unusual history and ethnic and religious makeup all contributed to Sarawak's distinctiveness in the context of Malaysia as a whole. The dominant conservation problem in the middle and upper Baram was logging too rapidly and too intensively, with too little planning and supervision. Meanwhile, too few areas were being set aside intact for maintaining biodiversity, for supporting local life styles, and for protecting the environment in general. Shifting cultivation was a limited threat because of the sparse and rather sedentary human population, while hunting was only a problem in relation to a few rare species such as rhinos. The main response to the logging problem in the Baram was to advocate a more comprehensive and environmentally sympathetic spatial plan for the whole area, while drawing attention to the hidden costs of poor logging practices and the potential benefits of improving both planning and logging.

8.2.2 Cross River, 1989–1990

Like Sarawak, Cross River State possessed many traditional and forest-dependent people, as well as ethnolinguistic groups not found elsewhere in the country. The State was unique in southern Nigeria in its low overall population density, and in having extensive forests and few roads, the former being rich in endemic and other species. Problems here were more diverse than in Sarawak. Commercial logging was a more important issue in the Oban than in the Okwangwo Division, to which the attempted solution was to provide technical assistance to the forestry industry, while highlighting the alternative benefits from donor support to the National Park.

Encroachment by small-scale farmers was an issue in both the Oban and Okwangwo Divisions, where help by the project to improve land use was to be exchanged for compliance by local people with Park boundaries. Pressure to expand plantations in the Oban Division was resisted by emphasizing the benefits of the Park. Hunting of threatened species was a serious problem in both Divisions, and was to be reduced through enforcement and public relations. Part of the response to hunting and gathering forest products in the larger Oban Division also involved conditional access to traditional use zones, and a partnership in policing them between the project and the local people.

8.2.3 Hainan, 1991

Hainan Province, relative to the rest of southern China, had significant numbers of traditional shifting cultivators, a low population density in the hills, and little urban and industrial development. The island also possessed very high levels of species richness and endemism. Nature reserves in Hainan were under threat from logging, encroachment, hunting, and overharvesting of forest products. These could be reduced through enforcement, but there were two other strategic problems. The first was that previous decisions on land use had caused great reduction and fragmentation of the area under natural forest. Meanwhile, rapid economic growth was being guided by plans which paid little attention to environmental concerns. Since it was not feasible to reverse the past conversion of natural forests to plantations, the approach used was to encourage linking of existing reserves and surviving forests in a few areas. New landscape management concepts designed to promote genetic continuity and rehabilitation of the environment were introduced. These were innovative in the Chinese context because they explicitly relied on planning and management with the participation of all groups affected in each area.

8.2.4 Siberut and Flores, 1992

Conservation problems in Siberut revolved primarily around excessive log-ging, as they did in Sarawak, but further aggravated by the wetter climate, by additional over-harvesting of non-timber forest products, and by various attempted resettlement and plantation development projects. They imply a failure of planning, as use of the island in these ways was clearly inappropri-ate from the point of view of local, forest-dependent people and native biodi-versity. Alternative development plans had been produced for Siberut in 1980–2, just as they had for the Baram area of Sarawak in 1984, but with equally little effect because the broader political context was not sufficiently supportive at the time.

The possibility of concessional financing for biodiversity conservation in Siberut seemed to offer a solution to the island's overall conservation prob-lem. This may only have worked in 1992, however, because the value of the island's remaining timber reserves had been greatly reduced during the pre-vious 20 years. It is also possible that Indonesia's increasing recognition of the value of biodiversity may have led to reduced pressure on the island within a few years.

Issues were simpler in the Ruteng area of Flores, where fuelwood collec-tion and small-scale agricultural encroachment by a small number of people were the main pressures on the reserve concerned. Both of these problems could potentially be relieved, and ultimately solved in combination with com-pensatory work, through enforcement and public relations efforts. The Ruteng forests were important for biodiversity because they were among the few remaining fragments of intact forest in Nusa Tenggara.

8.2.5 Makiling, 1993

Official attitudes to unofficial participation by local people were more posi-tive in the Philippines in 1993 than in China in 1991 or Indonesia in 1992. The main problem in the Makiling Reserve was that land ownership had become very confused due to settlement and farming by up to a thousand families. This had occurred in a small forest which was uniquely valuable as a biodiversity refuge, and threatened to destroy it entirely. The greatest need was for a sense of purpose by the university responsible for the Reserve in negotiating a solution with resident farmers. This was envisioned to mean involving them formally in saving the remaining forest, in return for granting them secure tenure over their existing farms. This was to be achieved in the context of a broader arrangement to manage the Reserve as a whole in partnership with farmers and other local stakeholders.

8.2.6 Costa Rica, 1992–1993

Costa Rica was attempting a comprehensive national response to conservation problems which had included high rates of deforestation throughout the 1970s and 1980s. This process involved general reform and rationalization of the nature reserve system, decentralization and devolution of management authority for reserves to local people, and exploration of new methods to finance conservation sustainably. The latter included the inventory and prospecting arrangements made by the National Biodiversity Institute (INBio). The total effect was to provide a dramatic model for biodiversity conservation which could be used to inspire officials from other countries, and to set them thinking of new ways forward which might apply in their own circumstances.

The Brunca Region of Costa Rica involved a diverse, complex set of conservation issues. They included ranching, fire, agricultural plantations, and infrastructure projects including dams, roads, and pipelines. A national response was well underway, through INBio and the various agencies of government responsible for protected areas and rural development, and locally through the efforts of several NGOs. Legislative reform was needed to release funding to Conservation Areas, but this appeared likely to occur in view of favourable events at a national political level. The main problem in the Brunca, which may have justified temporary external help, was a lack of co-ordination and common focus by the various groups interested in and active in the region.

8.2.7 Irian Jaya, 1994

Threats to the environment of the immense territory of Irian Jaya were diverse. They included road construction, industrial exploitation of biological and mineral resources, conversion of natural ecosystems, and the arrival of large numbers of settlers from other parts of the country. These implied that the government should be helped to supervise the development process more carefully, and also created a need for a flexible, site-specific response to problems arising in each location. An assumption was that government would wish to avoid unnecessary damage to the environment. It was therefore proposed to use maps and databases on Irian Jaya's biological and cultural diversity, to help all parties comply with spatial plans and EIA/AMDAL procedures, and thereby to avoid conflict with development projects. Policy changes in 1993–4 encouraged work around priority reserves to focus on helping local people to agree, mark and map the boundaries of land-holdings, and designing community projects to support reserve management. This two-tier strategy was to be combined with environmental education for all Irian Jaya residents, whether aboriginal or newly arrived.

8.3 Themes of the case studies

8.3.1 Economic values

The projects described in Chapters 2–7 involved partial solutions to complex and dynamic sets of problems, in which various themes were explored, any of which might be relevant to other conservation problems in the future. One recurring theme was that of trying to change people's perception of the economic value of renewable natural resources. The early stage of work in Sarawak, for example, was aimed at making a case for investment in wildlife management. It was shown that wild meat was worth many millions of dollars each year to Sarawak's economy, and that if this resource was lost it would either have to be replaced from domestic sources or rural people would suffer a loss of dietary quality. It was argued that investment in the regulation of hunting and the protection of wildlife habitats would be more efficient than meeting the cost of replacing wild meat, or accepting the political and humanitarian costs of widespread malnutrition.

A similar argument was used for other non-timber forest products in Sarawak, starting with rattans and tourism benefits. It was later broadened to include a range of values which might be obtained through the commercial development of biodiversity. This logic led in 1988–9 to proposals for a Natural Products Institute (NPI) in Sarawak, and later for Cross River Bioresources Limited (CRBL) in Nigeria. The aim was to try to help government officials to see rain forests as 'much more than stocks of wood' (Myers, 1988), a line of advocacy which was new at that time and remains compelling.

The idea that biodiversity has an important commercial dimension was greatly strengthened by the example of INBio in Costa Rica. The reasoning behind NPI in Sarawak and CRBL in Nigeria had been developed independently in Costa Rica, but with fuller attention to such details as the precise methods of collecting and using data, and the contracts needed to benefit all parties while helping to reward conservation investment. The Costa Rican experience also led towards the broader issues of how and why to decentralize the management of nature reserves. At a practical level, it was found that exchanging officials between Costa Rica and other countries could help to change official attitudes to biodiversity. In Indonesia, in particular, exposure to INBio in 1992 may have encouraged creation of an Indonesian Biodiversity Foundation in 1994, an organization with similar aims but adapted by its Indonesian designers to local needs and circumstances.

8.3.2 Participation and accountability

Advocacy based on actual or potential economic values can strike at the root cause of much planning, policy, and institutional failure. Even when alerted to the economic arguments in favour of conservation, however, individual officials may be unable to act on them. This is especially so where concentrations of wealth and power are combined with a lack of public accountability, since there may be an élite whose self-interested decisions can overrule those of environmentally aware civil servants or local people.

This is why another recurring theme in Chapters 2–7 was that of increasing public participation and accountability in project design. This began from a low level in Sarawak, where some politicians and senior officials saw research based on asking rural people about hunting as unreliable, and a few saw proposals to strengthen local tenure of forest resources and wild meat supply as subversive. The research issue was then eclipsed by a debate about logging which became intensely political. Nevertheless, public support for communal forest reserves, community rattan plantings and national parks (such as Pulong Tau or 'Our Forest') made an impression, and public opinion was later canvassed more formally through the State Legislative Assembly's study on wildlife issues in Sarawak.

The role of local participation became more overt through the village liaison assistant scheme in Nigeria, the role of which was to maintain a continuous, two-way flow of information between local people and project managers. From the effect of creating a smaller Cross River State, it was learned that conservation is more likely to happen when the people most likely to benefit from it are the ones who decide whether or not to do it. Local involvement was then built into plans to manage the Jianfengling and Wangxia protected landscapes in China, the Makiling Forest Reserve in the Philippines, and several reserves in Indonesia. Especially in Nigeria, China, and Indonesia (i.e. in 1990–2), these projects sought to reconcile the needs of local people with those of central governments, large banks, and consulting firms.

This led to compromises in each project design (Caldecott, 1992b). Firstly, committees were to be used to oversee the project and to co-ordinate concerned institutions, while absorbing technical advice from consultants. These committees were to be consistent with existing administrative arrangements in each country, and were to be established at the central, provincial, and district levels in Indonesia, the provincial and 'protected landscape' levels in China, and at the federal and state levels in Nigeria. Their role would be to ensure horizontal communication among interested groups at each level, while also allowing decisions which could not be made locally to be passed to higher or more central levels for discussion.

These oversight committees were to interact with the project level through local or locally acceptable NGOs, which were to act as contractors to government and to the donor agencies involved in each case. The rationale for increasing the role of NGOs was based on a review by Wells, Brandon, & Hannah (1992), which had concluded:

- that NGOs are often the source of new ideas and project concepts;
- that they often act as institutional bridges between governments and donor agencies, and between projects and the local people within project areas;
- that they tend to be sensitive to local needs, flexible, and inclined to find site-specific solutions to problems;
- that they tend to promote a consultative, 'bottom-up' style of planning and implementation; and
- that they often have few resources, and respond well to opportunities to form partnerships among themselves and with governments and donor agencies.

The role of NGO contractors was to undertake small-scale tasks which called for working closely with local people. These included promoting literacy, primary health care and family planning, environmental education and conservation awareness, social forestry and agroforestry extension, organizing and training tourist guides, introducing new husbandry techniques, and assisting in defining local development priorities. Detailed criteria were worked out to guide the selection of NGOs by donor agencies and governments before they could be awarded such contracts (Box 8.1). The main effect of this arrangement was that NGOs were given a formal role in project implementation, rather than being excluded entirely as had previously been common in the countries concerned. This tended to shift responsibility for proposing project development initiatives away from government consultants and towards plans prepared with local community involvement. The main problem was that NGOs draw their strength from members who are committed to priorities which often differ from those of governments. Trying to make NGOs work as government contractors therefore posed the risk to the NGOs of losing their sense of identity and purpose, and of compromising their relationship with local people.

Box 8.1 Criteria for NGO involvement as contractors in conservation projects (adapted from Caldecott, 1992b; ADB, 1992b)

Selection should be limited to those NGOs:

- which are registered in accordance with the law;
- which can provide audited financial statements;
- which have been in operation for at least two years;
- which have worked with local people at a similar stage of socioeconomic development as those in the project area;
- which are able to field an adequate number of qualified, experienced, and trained staff; and
- which are willing to establish an office with full-time supervisory and support staff in the project area.

The implementing agency should invite contract proposals covering:

- the experience of the NGO in community organization among rural groups;
- the results achieved in the socioeconomic development of rural groups;
- the extent of work in rural areas;
- the profile of field and administrative staff; and
- the proposed work plan and financial plan for the assignment.

Evaluation of the NGO's performance should be based on:

- the quality of community development plans produced;
- the quality, nature, and quantity of other work achieved;
- the level of income achieved or maintained per beneficiary family; and
- the degree of co-operation by the community, the extent of self reliance, and the degree of community support for conservation objectives.

The NGO should be required to produce reports, including:

- an annual work plan and financial plan;
- monthly progress reports on activities and the achievement of targets; and
- quarterly and annual progress reports.

Financing arrangements should include:

- the financing of activities, staffing and overheads according to agreed rates;

- an advance of funds for up to three months' work, on an imprest account basis, with replenishment according to the annual work plan and financing plan; and
- strict accounts of project expenditures, to be submitted periodically and annually.

There should be preference in favour of those NGOs:

- with adequate technical capability in environmental matters;
- with a broadly based membership and strong impact at the grass-roots level;
- with no pending cases at law against them or their key officers;
- with well-defined areas of operation and a record of good performance and reputation both on the ground and among government officials;
- which are financially viable;
- which have adequate management and absorptive capabilities, shown by yearly expenditure and achievement, and with administrative costs below about 25% of total costs;
- which are able to work with other local or non-local NGOs or which, if not themselves local, have local counterparts which they are working to strengthen; and
- which encourage direct participation of local people in their operations and in the direct receipt of benefits.

Involving NGOs formally in project design and implementation was a cautious step in the direction of open planning and devolved management of nature reserves and projects. This was part of an overall trend towards reducing the chance of central groups causing damage, by increasing both the power and the accountability of local ones. It was carried further in Costa Rica and, by different means, in Irian Jaya.

In Costa Rica, local conservationists saw participation as being fundamentally linked to biodiversity valuation. Their aim was to create a consensus that ecosystem protection was a local rather than a central responsibility, and to ensure that local groups benefited from doing it. Thus, a central data-management agency (INBio), and a national biodiversity inventory, were to complement and help finance conservation areas under local management. This approach resulted in a creative and practical synthesis of local and national interests. Costa Rica began planning for these reforms with a detailed national study of the issues. This study was carried out with strong political support, and it took into account both scientific knowledge and the country's own administrative and other constraints. Opportunities for improvement were clearly identified, and necessary measures discussed, agreed upon, and systematically put into effect.

In Irian Jaya, local conservationists were primarily concerned with safeguarding the way of life of people who depended on continued access to intact natural environments. Secure tenure over resources was seen as a precondition for local participation in conservation. The aim of such involvement was to make it easier for government to comply with its own spatial plans for the reserve system. Thus, the emphasis was on community mapping of reserve boundaries and land-holdings, and local co-management schemes for reserve resources. This needed to be integrated with government planning and impact management procedures, and policy changes occurred in 1994 which encouraged this.

8.4 Feasibility studies

Most large, long-term projects will need a specific research and planning phase, the results of which will be reviewed by donors and governments before further investment is committed by them. The need for this work is partly practical, and partly to reassure donor agencies and governments that the project is sound and adapted to their own policies and procedures. Such a study may involve a number of specialist consultants, but in view of their cost and short-term role, these consultants should be used cautiously and preferably as a complement to full-time conservation staff. Some guidelines for using consultants are given in Box 8.2.

Box 8.2 Using consultants in feasibility studies and projects
(adapted from Caldecott, 1991d)

(a) Establish the exact nature of the problem to be solved.
(b) Establish which questions need to be answered before you can formulate an opinion on that problem.
(c) Determine how these questions may best be formulated to direct attention to the critical issues.
(d) Consider how those questions might be answered, and whether a specialist consultant is really necessary in light of all the options.
(e) Specify the terms of reference (ToR) based on (d).
(f) Determine the plan and duration of work based on the ToR.
(g) Decide how much the answers to the questions in (b) are worth to you, relative to available funds and the schedule of the project.
(h) Identify a consultant who you believe will be able to obtain the answers in the period and for the fee which you can afford.
(i) Before signing the contract, reconsider everything in terms of whether you will have confidence in the answers provided (e.g. whether the questions

can really be answered in the prescribed period, or at all).
(j) Make sure that the consultant understands exactly what you want to know before work is started.
(k) Discuss progress with the consultant often during the study, ask a lot of questions, and do not accept what you do not understand.
(l) If the consultant cannot provide a final report before leaving the project area, make sure that you obtain at least a written summary of all findings and recommendations (and the consultant's contact numbers and address), and that you have the opportunity to ask questions.

Examples of the range of subjects covered in a number of feasibility studies were given in Chapters 2–4, and a model of the kind of budget that might result from such a study is given in Box 8.3. The feasibility study will have several roles, including helping to design the project, helping to obtain resources for it, and helping to establish it in the project area. These roles may relate to different priorities among the various groups involved in the project. An NGO might make a long-term commitment to a project, for example, but may ask another agency to fund the feasibility study. The latter will produce a report which the NGO can use to raise funds from various sources for further work. This work may be done by the NGO itself, or by others such as the consultants and staff of official donor agencies and governments. In this way, an NGO can turn its intellectual commitment to a project into a large and expensive field programme, without actually spending much of its own money.

Box 8.3 A model project budget resulting from a feasibility study, described by budget line (adapted from Caldecott, Oates, & Ruitenbeek, 1990)

1. *Vehicles & equipment.* Assume that these are purchased in Year 1 and replaced in Year 5.
2. *Vehicle running costs.* Assume a flat rate per vehicle-year for saloon, minibus and 4WD (times 200% for a truck or 20% for a motorcycle), for fuel, spares, repairs, and insurance. State the share of costs to be accounted in the local currency.
3. *Recurrent non-staff costs.* Include office, maintenance, and operational costs not specified elsewhere.
4. *Local Staff.* Include salary, allowances, and operational costs per person-year by grade (e.g. senior management/technical specialists; middle management/technical assistants; and clerical support staff/drivers).
5. *In-country training scholarships.* Allow for in-country training involving a suitable number of person-years per year at an assumed all-in cost as indicated by the donor, which should be phased in over Years 1–4.

6. *Building and refurbishment costs.* New or refurbished buildings entail construction, electrical, plumbing, furnishing, and other costs early in the project, plus a recurrent spending commitment, which can be minimized by specifying, for example, designs using natural ventilation and fans, rather than air-conditioning in tropical locations.
7. *Technical Assistance Staff.* Include long-term international appointments at an assumed all-in cost as indicated by the donor.
8. *Consultancies.* Include short-term international appointments at an assumed all-in cost as indicated by the donor.
9. *Aerial survey.* Allow for reconnaissance and investigative missions.
10. *Regional training scholarships.* Allow for training within the region involving a suitable number of person-years per year.
11. *Overseas training scholarships.* Allow for training outside the region involving a suitable number of person-years per year.
12. *10% Price contingency.* This will represent global inflation and local inflation not balanced by exchange rate changes. The budget may otherwise be given in uninflated units, although this will make it hard for many people to understand.
13. *10% Physical contingency.*

Establishing a project involves defining policies, strategies, targets and schedules, as well as hiring and training staff, obtaining accommodation, vehicles and equipment, and making arrangements to administer the project and to meet its day-to-day requirements for money, supplies and services. Some of the practical implications for project managers are discussed by van Lavieren (1983), Pratt & Boyden (1985) and MacKinnon *et al.* (1986), and a few of them are listed in Box 8.4.

Box 8.4 Practical project management issues (adapted from Caldecott, 1991d).

Banking and foreign exchange. Consider: opportunities to use and consequences of using official and unofficial exchange rates, travellers' cheques, and credit cards; establishing a local foreign currency account; the speed and reliability of transfers from abroad and local banking services.

Local staff. Consider: legal obligations of employers, availability of staff by educational standard and ethnolinguistic group, grades and job titles, basic wages, overtime, allowances (e.g. for transport, housing, inflation, field subsistence), leave, loans, holiday and seasonal bonuses, insurance, medical care, probation, disciplinary and dismissal procedures, and the hours and days of work. Note that there may be tension among local groups which may make it easier to hire an outsider than a local person, at least for a while or to achieve a particular objective.

International freight and communications. Consider: the best ways to deliver goods, supplies, and equipment (e.g. by land, by sea, or as air-freight or excess baggage), and to communicate locally and overseas (e.g. by telephone, fax, computer link (e-mail), telex, radio, courier, or mail).

Choice and maintenance of vehicles. Consider: terrain and climate; local availability of gasoline (petrol) and fuel oil (diesel); optional specifications (e.g. regroovable tyres, halogen headlights, high-level air intake, oil cooler, winch, and roof-rack); training drivers (e.g. in driving technique, loading, and care of the vehicle); frequent and regular servicing (e.g. with oil and filter changes every month or every 5000 km); local availability and quality of spare parts and tools and the need to import them; local availability of mechanical expertise and the need to train mechanics.

Computer equipment. Consider: choice of portable computer relative to anticipated needs of compatibility and networking, and tasks such as managing databases, text, and spreadsheets; protection against power fluctuations, damp, fungus, dust, insects, and computer viruses; local availability of software and hardware expertise.

Health hazards. Consider: training and materials needed to avoid specific local diseases; consequences of local diseases which cannot be avoided; how to manage chronic illness and stress; insurance policies; emergency evacuation procedures; local availability of medical and surgical support, equipment, and medications.

8.5 Making projects local

Many options were mentioned in Chapters 2–7 for using and protecting living resources both inside and outside nature reserves. Further details of these options are given in Chapters 9 and 10. Here, the aim is to guide the process of making an initial choice of how to respond to a set of specific problems in a particular location. This choice should reflect local conditions in the project area, and will need to be made as a first step either in assessing or in designing a conservation project. The rationale is similar to that of 'pre-planning' (Bridger & Winpenny, 1987), in that the aim is to see if the project makes sense before any further work is undertaken.

A first step is to ask a series of questions about the potential project area, the first group of which concern whether there is really a problem which must be solved. The answers may be fairly clear from existing reports, but these may be unreliable and circumstances in the field may have changed since they were prepared. A preliminary visit to the proposed project area will therefore be required. Questions at this stage will focus on the size, status, and ecological condition of the reserve, and what kind of threats there may be to it. The most common danger signs include:

- if the reserve has many people living nearby, who may require farmland and other resources;

- if the reserve is about to be logged, polluted, cleared for plantations, mined, or built upon;
- if the reserve is about to be affected by infrastructure which will increase access, settlement and use patterns; or
- if the reserve's wild plants and animals are being exploited commercially.

The next questions concern the nature of the area to be protected. How is it faring in general? Is it large or small? Are the ecosystem types within it unique, rare, or common? Is the area predominantly intact or damaged, and if damaged by how much and in what ways? If artificial ecosystems are already widespread within it, why, when, and how did this happen? Is it possible to imagine those areas being rehabilitated, for example by being protected from fire, grazing, or dynamite-fishing long enough to be recolonized by native species? Are there natural ecosystems nearby, and if so how close and how intact are they? Is the ground vegetation cover predominantly intact in the project area, and are there signs of soil erosion?

The next group of questions will involve assessing the urgency and scale of the problems affecting the reserve, and the actions and resources needed to solve them. If a reserve is subject to small-scale encroachment, or hunting, fishing, and gathering, then local solutions are likely to be feasible. To establish its circumstances, it is necessary to know the answers to questions such as the following. How fast is the population growing? How many people seek new land each year? What social controls already exist on access to and use of natural resources, including land and fishing grounds? Solutions are likely to be easier to find where local populations are culturally homogeneous and have lived in a place for a long time. This is because they tend to have established arrangements to regulate access to natural resources, and to avoid or resolve resource-use conflicts among members of the community. A project would normally try to work with the existing leadership of such a population, and to use local procedures to resolve problems wherever possible.

If local groups are diverse and potentially hostile, however, then a different approach will be needed. This should be aimed at reducing tensions by helping people to talk with each other, and helping them to negotiate peaceful solutions to conflicts of interest that may exist. This may involve using group facilitation, negotiation, and conflict management techniques which are discussed further in Chapter 11. A common forum or management committee may help develop a collective approach to the project. There may also be a need to divide the project area into zones for exclusive use by different groups, including partition of buffer zones or traditional use zones where wild species can be harvested.

Having clearly outlined the problem, the next questions concern whether the main issues can be resolved locally, or whether a shift in policy is required which local people and institutions cannot achieve by themselves. Often both local and non-local responses will be necessary, and these will also need to be coordinated. Making a project local implies that each community should be able to express its conception of the environment in which the community lives, and the options, priorities and need for assistance in improving their use of that environment. Planning and environmental education should therefore be seen as inseparably linked.

A project should also aim to employ local people, either directly or through local NGOs, and should especially seek to employ and train them to work in their own communities. Local staff well supported as residents within the project area will be effective and committed members of the project. As shown in the case of parataxonomists in Costa Rica (Chapter 6), training of local people in biology and taxonomic techniques also allows them to participate in biodiversity inventories and increases their impact on local awareness. Life styles of local people reflect prolonged experience in the project's environment, and local knowledge of how to use resources can greatly assist the project's design.

This applies both to knowledge of local species and cultivars in farming systems (Chapter 10), and to traditional arrangements which may ensure that resources are used more equitably and/or more sustainably than modern alternatives. An important starting-point for a conservation project, therefore, will be to understand how local people make a living in the project area, and how this life style affects the environment to be conserved. A head start can be obtained by employing local people, who will begin with background knowledge and personal contacts which an outsider could take many years to develop. The same logic applies to involving local people in planning, supervising, and overseeing the project as a whole.

8.6 Making projects complete

A short field survey will usually show where and how the strongest ecological and economic factors affect a reserve. The aim is then to focus on the most relevant geographical area to be covered by the project, which will include the reserve itself, nearby natural systems and settled areas, and the people who live there and use resources in the reserve. Environments 'upstream' from this region, whether by river or marine current, are also of interest, as are any policies, plans, and investments elsewhere which may affect the project area. A project may fail if it is designed to respond to local problems, when those problems originate far away, or if future threats such as population growth are not anticipated. A major infrastructure investment might be planned by another agency near or within a reserve, for example, but neither the project nor the local people may hear about it before budgets have been committed or work started. If it is thought that underlying threats can be controlled only through massive investment or long-term policy change, it may instead be reasonable to decide to spend scarce resources elsewhere.

When making such decisions, it should be recognized that there may be merit in opposing the root cause of a major problem, and in working to reduce damage while using the struggle to raise awareness of conservation issues. The outcome may be changed just by trying to make a project work, and enthusiasm and commitment can achieve surprising results. Work on a troubled species, for example, may attract media and political attention, leading to new funding or new policies. Otherwise, a community may campaign against commercial logging, and find itself unexpectedly with powerful allies and a new nature reserve. It is thus not always possible to predict what will work and what will not, and extraordinary outcomes may result from the willingness of conservationists to take risks.

Because of the analogy with the way water can carry impacts downstream into a reserve, the process of looking for distant causes of actual or potential problems can be called 'upstream analysis'. At a literal level, this is implicit to the idea that projects should encompass whole ecological units, for example entire estuarine systems including all inflowing rivers, or lowlands and slopes up to the highest watershed above them. Upstream analysis, however, also involves trying to understand the implications for the project area of policies, plans, and administrative arrangements devised by people who have no direct contact with the project area.

These may include the effects of macroeconomic policy. An overvalued exchange rate, for example, may give some protection to forests because it has the effect of taxing agriculture, thereby reducing the demand for new lands to be placed under cultivation (S. Rosenthal, personal communication, April 1995). If there is a devaluation, pressure on the forest area may increase in line with the agricultural supply response to be expected from such an action. Such an effect was seen in Nigeria as a result of Federal structural adjustment policies after the mid-1980s (Chapter 3). A devaluation might also provide opportunities for conservation, since, depending on the choices made and local management capacity, it can be used to promote tourism, to raise stumpage in forestry, and to attract helpful investments.

If a reserve is threatened by hunting and encroachment, a project might seek to relieve hunting through policing and public-relations work, perhaps with compensation for lost food and traditional revenues. It might also try to relieve land hunger by improving nearby agriculture, by sharing benefits from the reserve locally, and through environmental education. A look beyond the immediate vicinity, however, may reveal factors which make it hard to achieve a purely local solution. Nearby forests may have been taken over by government or a private corporation, closing them to local people and depriving them of opportunities to hunt and gather there. Logging or commercial agriculture in those forests may have caused erosion or flooding which has damaged fish stocks or crops. These events may have caused or aggravated the problems which the project is trying to solve, but they may depend on decisions or policies made in towns far away.

An example is provided by conservation efforts in the Hadejia-Jama'are floodplain in north-eastern Nigeria (Adams, 1987; Kimmage & Adams, 1992). This floodplain is a very productive seasonal fishing, grazing, and farming resource for local people, as well as being an important site for migrating birds. Annual floods in the area also recharge an aquifer which underlies a much larger area of the Chad Basin, and which sustains agriculture and settlements far from the rivers themselves. The latter drain parts of Kano and Bauchi States, and flow north-eastwards through Jigawa and Yobe States into Lake Chad.

Community-based conservation projects in the floodplain have been undertaken by BirdLife International (ICBP), the World Conservation Union (IUCN), and the European Union through its North-East Arid Zone Development Project (Barbier, Adams, & Kimmage, 1991; Hollis, Adams, & Aminu Kano, 1993). These efforts were undermined by dams and irrigation schemes in Kano and Bauchi, which prevented the annual flooding on which the ecology of the floodplain depended. The conservation projects therefore sought to document the economic consequences of upstream dams for the Hadejia-Jama'are system and the Chad Aquifer. They then used this information to try to influence decisions on the management of water resources in north-eastern Nigeria as a whole. They also explored alternatives, such as that of restoring annual flooding by releasing water from the Challawa Gorge Dam in Kano. The donor groups concerned thus decided to continue local investment in the project area, while also seeking solutions to strategic threats originating elsewhere.

Other issues arise from conflicts between the interests of large national or trans-national corporations and those of conservationists and local people in the areas affected by them. An example is that of mining interests in the Lorentz reserve in Irian Jaya (Chapter 7), and increasing numbers of such conflicts exist elsewhere. The trans-nationals involved possess abundant financial resources, but are poorly regulated by national authorities, since they can move funds freely among their national subsidiaries. They are a challenge for conservationists, since they can bring very strong pressure to bear on the authorities responsible for protecting nature reserves. There is an urgent need for conservation groups to find ways to influence trans-nationals, based on helping them to accept their moral and practical responsibilities towards local societies and environments (N. J. Ashton-Jones, personal communication, March 1995), on their role in sustainable development (IISD, 1994a, 1994b), and on strengthening the capacity of local people to resist damage to their environments through mobilization and national decentralization (Chapters 5 & 6).

8.7 Keeping projects solvent

Most wild species will be safe only to the extent that the people who conserve them feel themselves to be rewarded by doing so. Rewards can take many forms, not all of which can be valued in financial terms. Money need not always be involved for all people in all places, but is usually an important factor for many. If it costs people more money to conserve than they gain from conserving, then it can be hard to argue that conservation is in their interests. A project may be able to do so successfully, if people are prepared to accept non-monetary rewards in exchange for monetary costs, but their willingness to do so depends on their attitudes and life-style priorities. A project which keeps real monetary costs to people to a minimum, and real monetary gain to a maximum, will thus usually be safer than another project which does not, provided that the people who benefit are the ones who decide the future of the resource to be conserved.

Conservation projects should be designed to ensure durable financing if possible, and there are many ways to achieve this, including those listed in Box 8.5. Whatever the source of funds for a conservation project, they should be used by the project as a whole, and spent under the authority of a local authority in which all stakeholders are represented. The first will help prevent certain components of the project being promoted out of balance with others, while the second will maintain public support and accountability. Political and administrative problems can arise if some reserves become much wealthier than others, due to luck or better management. This arose, for example, in relation to the well-endowed Guanacaste Conservation Area in Costa Rica, compared with reserves in the Brunca Region of the same country (Chapter 6). The consensus which had emerged in Costa Rica by late 1993 was that all parts of the national reserve system should share in all revenues according to their needs, but that individual areas should be able to retain enough locally generated funds to reward local initiative and success. The details of revenue allocation are likely to be a permanent subject of debate in any such arrangement.

Box 8.5 Options for funding nature reserves

External subsidies. Governments may agree to make perpetual payments, in exchange for national or global benefits from the reserve. How useful this is will depend on the long-term reliability of the external donor, which is often low if funds are allocated from general taxes in competition with other government priorities.

Earmarked payments. Financing is more reliable when the funds are raised by levies earmarked for the account of the national conservation system. Where nature reserves and tourism are seen to go together, such taxes might be raised from arriving tourists, or from hotel services, luxury goods, or tours booked locally. This will at least ensure that each reserve will share funds from a system controlled by people whose jobs depend on the survival of the national system of nature reserves.

Funds raised locally and retained. Charges may be made to enter, film, or conduct research in the reserve, or to stay overnight there. Taxes may be charged or profits made on the sale of local goods and services, such as on bottled water and honey or on guide and translation fees. These may provide very reliable sources of finance for a single reserve, if the funds are retained locally and used to underwrite its management budget.

Charges for ecological services. It may be possible to arrange for beneficiary groups to pay for ecological services provided by a reserve, such as when farms and towns downstream are charged for water supplies and flood prevention services. A drawback is that many such beneficiaries may be unwilling or unable to make such payments.

Charges for collecting specimens. Reserves can also sell permits to collect wild species, such as ornamental plants or hunted animals. This is linked to the use of contracts to allow biodiversity prospecting, which provide for up-front payments and royalties on any commercial products that may be discovered.

Investment portfolios or endowments. Investments of an ethical nature may be made to ensure that a reserve's revenues will always match or exceed its financing needs, thus allowing long-term planning. Endowments also offer a way to absorb and store surplus funds, such as bequests or one-off grants, against future need. It is hard to manage investments to yield a steady income, however, and there are no risk-free investment strategies. Contingency planning, hedging of investments, and professional help will be needed to reduce risks and maximize safe returns.

Offset agreements. Other funding possibilities include power companies paying to protect forests or coral reefs to offset their carbon emissions. A related concept is to use reciprocal taxation agreements, which may allow a company to offset tax liabilities in its home country by an amount equivalent to that donated to the reserve. It might also be possible for a reserve to insist upon conditions before allowing an investment in an area. A licence to build a beach hotel near a reserve, for example, might be exchanged for help in protecting reefs and marine wildlife.

8.8 Valuing projects

Economic valuation techniques are important for at least two reasons. Firstly, they can be used to help reassure and persuade donors and governments that conservation investments are worthwhile (Ruitenbeek, 1990b; Chapter 11). Secondly, they can help project designers to avoid mistakes which may reduce a project's viability (Bridger & Winpenny, 1987; Munasinghe, 1993; Turner, Pearce, & Bateman, 1994). Projects have costs as well as benefits, which must be measured in a common currency if they are to be compared. A project's costs tend to be the easier to measure, and reflect the lost opportunity to spend resources in ways other than on the project. A project's benefits are often more difficult to measure, since in the case of conservation projects they mostly reflect the economic value of the environmental damage that is avoided by undertaking the project.

The economic analysis of a project should encompass a number of steps, of which the most important is to make sure that all costs and benefits are taken into account in the study (CEC, 1993). This should apply even to those for which it seems unlikely that a monetary figure can be obtained. Thereafter, valuation will involve a series of other steps:

- Firstly, assuming all options for action are feasible, they will all need to be defined, including the option of doing nothing at all.
- Secondly, a full balance-sheet of benefits and costs will be needed for the various options, in quantitative and then in monetary form. The benefits will include both the positive benefits of action and the costs of doing nothing which have been avoided by action. The costs will be based on

achieving whatever conservation targets have been set by the most cost-effective set of actions.

- Thirdly, the costs and benefits will need to be compared for each feasible option, as a basis for formal cost-benefit analysis.

A critical issue in valuing environmental costs and benefits is the choice of methods to measure or estimate them (see also Chapter 11). Methods based on real market values include:

- effects on production (e.g. of crops, fish, wildlife, and timber);
- losses of earnings (e.g. from health effects);
- preventative expenditure (e.g. by training loggers to fell trees more carefully, by building better logging roads, or by patrolling to prevent over-fishing); and
- replacement cost (e.g. by replacing lost protein in hunters' or fishers' diets from domestic sources).

It is also possible to use an imaginary market to establish values, in which people may be asked to make a contingent valuation: to express a willingness to pay for something or to be paid to accept its loss. Other options include hedonic pricing, by looking at real markets which shed light on the value of environmental attributes which are not traded themselves. Thus property values can reflect environmental quality, and time or money spent in travelling to a site can also help to value it, whether the site is used for recreation or for collecting fuelwood or water. Another form of hedonic pricing is product substitution, such as where the price of domestic meat is used to value wild meat or that of kerosene is used to value fuelwood. If the market price is distorted by government subsidies or other factors, then a corrected shadow price may be used (Munasinghe, 1993). All these methods of valuation are often used to help governments decide whether or not to support projects, and conservationists should therefore know how to use economic analyses in support of their proposals.

An important issue for valuation is the discount rate applied to the predicted streams of economic costs and benefits, and used to weight their value according to when they are incurred or received. The assumption underlying a discount rate is that benefits, resources, or utility gained or lost are worth more if this happens sooner rather than later. There are good reasons for this, especially that people know they will die but do not know when, that people prefer receiving real benefits themselves rather than risk those benefits going to others later, and that money obtained (or not lost) now can be invested to increase later security.

Discounting makes short-term economic benefits more attractive than long-term ones, and the strength of this effect depends on the discount rate used (Munasinghe, 1994). Applied to a choice of ecosystem management investments, for example, a high discount rate would mean that cash obtained from selling timber or wild meat may be valued more highly than avoiding later droughts and floods, or gaining future tourism revenues. Such discounting will also make immediate costs less attractive than later ones, thus discouraging people from investing in education or durable farming, for example, and encouraging them to accept the later costs of ignorance or soil erosion. It is thus hard to justify investments in sustainable development if high discount rates are used.

On the other hand, applying low discount rates to a whole economy would encourage all forms of investment, including those which may damage environments, and may also have the effect of trapping current generations in poverty while discriminating in favour of future ones (Munasinghe, 1994). Special economic analyses are therefore needed in the case of conservation projects. The European Commission's *Environment Manual* (CEC, 1993), for example, suggests that standard cost-benefit analysis can be improved by carefully distinguishing between long-term and short-term costs and benefits, and by giving special significance to non-monetary consequences, particularly irreversible ones. An analogous idea is that of applying a 'sustainability constraint' to investment, which would maximize current well-being without reducing that of future generations below current levels (Munasinghe, 1994). A reason for this effort is that, although people are mortal, human societies are less so, and should therefore value the future more highly than individuals are inclined to do.

9

Options for conservation

9.1 Introduction

Project areas are clearly bounded regions within countries, or at least have
to be defined and mapped as such in project documents. Each project area
will contain a number of actual or potential zones, based on features of terrain, ecology, or human use of resources. Conservation projects must address
the needs both of wild species and of people, so when they are designed the
focus will switch back and forth between the most important zone for biodiversity (the nature reserve), and the most important area for solving problems
(the surrounding area where people live). This chapter reviews some of the
options for managing the various zones which comprise nature reserves themselves. The next chapter complements this by focussing on the options for
managing the rest of the project area.

9.2 Choosing nature reserves

There is no single objective way to measure the importance of the various
components of biodiversity, or to choose priorities for managing them, since
these actions depend on the aims of those doing the measuring and choosing.
The aim of preserving genetic variability among crop plants, for example,
yields different priorities from the aim of preserving wildlife for viewing by
tourists. For the purposes of this book, however, it is assumed that the primary aim of a conservation project is to preserve as many species as possible
in the short to medium term, while seeking ways to resolve threats to them
in the long term. People who propose or design such projects should therefore
be guided by the distribution of native and particularly endemic species, and
should emphasize protection of areas which are rich in species and endemics.

The opportunity to choose where to create new nature reserves is quickly
being lost in most countries, as natural habitats are used or set aside for

other purposes. On the other hand, governments and others still have to make decisions about where limited resources should be invested, in order to achieve maximum benefit for conservation. These choices should also be influenced by the distribution of species-richness and endemism, even though many other factors may also be important. There is an abundant literature on the biological theory and practicalities of wildlife management as they affect the design of nature reserve systems (reviewed by Caughley, 1994). This draws on knowledge of the effect of environmental fluctuations on demography, and genetic factors such as heterozygosity and inbreeding, fitness, genetic drift, and their relationship to effective and minimum viable population size. As the scientific basis for conservation decisions becomes stronger, conservationists become better able to specify the survival requirements of wild species.

The minimum investment needed to avoid species extinctions in a country is that which is enough to ensure the safety of representative samples of all its distinct natural ecosystems, plus some duplicates to offset the risk of mistakes, local catastrophes, and other factors such as climate change, with some extra reserves for rare or endemic species which may not be included otherwise. This minimum investment may be hard to achieve because of a shortage of financial and/or other resources, in which case the shortage is a priority issue in itself. Assuming it can be resolved, choosing among possible priorities then demands understanding of the nature and intensity of each threat to each reserve, and also the options for controlling each threat including the cost of each option (WWF, 1995b). This calls for information, and for creative and flexible thinking.

If a reserve is threatened by a government-built road, for example, the cheapest solution may be to use official planning procedures. This might involve campaigning for an environmental impact assessment (EIA), which would provide an opportunity for public opinion and technical advice to exert influence on the decisions involved. If the same reserve is threatened by colonist farmers, however, the cheapest solution may be to clarify local land tenure. This will involve helping residents to agree and map land claims and reserve boundaries, which would have the effect of deterring new settlement into a landscape in which all parts are legally claimed and occupied by someone or some agency.

International agencies must also make choices about where to invest, and may wish to refer to the global distribution of species when they do so. This can lead to conflict with national groups, which may wish to set priorities according to national aims instead. Countries which possess most species-richness and endemism, for example, are listed in Box 9.1. To an international conservation agency, it may make sense to help these countries, rather than others, to comply with Articles 6, 7, and 8 of the Convention on Biological Diversity. This would mean that they would complete their national programmes for biodiversity management earlier than others, which should result in quicker and better conservation than if less species-rich countries were helped first. This approach would not be expected to appeal to nations which are either not on the list, or which are opposed to international agencies choosing priorities on their behalf.

Box 9.1 Countries or territories which possess most terrestrial species and most endemic species, in alphabetical order within each group (adapted from Caldec-ott *et al.*, 1994; based on species-richness and endemism among mammals, birds, amphibians, reptiles, swallowtail butterflies, and angiosperms).

Group 1 (the 25 most biodiverse countries): Argentina, Australia, Bolivia, Brazil, Cameroon, China, Colombia, Costa Rica, Ethiopia, Ecuador, India, Indonesia, Madagascar, Malaysia, México, Papua New Guinea, Perú, the Philippines, South Africa, Tanzania, USA, (ex-USSR), Venezuela, Vietnam and Zaire.

Group 2 (the 25 next-most biodiverse countries): Angola, Botswana, Cambodia, Central African Republic, Chile, Congo, Côte d'Ivoire, Cuba, Gabon, Ghana, Guatemala, Guyana, Iran, Kenya, Laos, Myanmar, Nigeria, Panamá, Paraguay, Sudan, Suriname, Thailand, Turkey, Uganda, and Zambia.

Group 3 (the 20 island countries with high endemism not in Groups 1 or 2): Comoros, Dominican Republic, Federated States of Micronesia, Fiji, French Polynesia, Haiti, Jamaica, Japan, Mauritius, New Caledonia, New Zealand, Palau, Puerto Rico, Sao Tomé and Principe, Seychelles, Solomon Islands, Sri Lanka, Taiwan, Vanuatu and Western Samoa.

There are at least three other ways for international agencies to choose where to put their conservation resources, although all may encounter the same kind of resistance from national groups. Firstly, they could be guided by the fact that about 221 endemic bird areas (EBAs) have been identified by BirdLife International (ICBP, 1992, 1995). About 26% of all bird species are restricted to these EBAs, which together amount to less than 5% of the world's land area. Since EBAs were defined so as to take into account their importance for other groups of terrestrial organisms as well as birds, the benefit from conserving EBAs is not limited to birds alone. Secondly, agencies could be guided by the locations of about 250 centres of plant diversity (WWF & IUCN, 1995). These are areas which are known to be rich in habitat types, species, endemics, and wild relatives of crop plants. Their protection would safeguard the majority of the world's wild plant species, as well as many other components of biodiversity.

Lastly, priorities for conservation have long been sought among the world's existing nature reserve system, and the results could guide international agencies in the future. Examples include reviews of reserve systems in the Afro-tropical, Indo-Malayan, and Oceanian realms (MacKinnon & MacKinnon, 1986a, 1986b, 1986c; Braatz *et al.*, 1992; SPREP, 1992; MacKinnon, 1994a). Many national studies have also been carried out, including environmental action plans, tropical forestry action plans, conservation strategies, and biodiversity country studies. This body of work can be used to focus attention on sites which are consistently identified as being of special importance for biodiversity.

Thus, there are lists for some countries of sites which have been repeatedly identified as priorities for protection if key components of biodiversity are to survive. Examples include 80–90 sites in Indonesia, which were shortlisted in successive studies by MacKinnon & Artha (1981–2), RePPProT (1990), BAPPENAS (1991), MoF & FAO (1991), and KLH (1992). Similarly, many important wetland sites have been identified in multi-country studies (e.g. Scott, 1989), and about 1000 locations worldwide have been listed under the Ramsar Convention as vital wetlands, or else are listed as Biosphere Reserves or natural World Heritage Sites. Locations which emerge consistently as priorities in such studies are compelling candidates for early conservation investment.

Priorities for marine biodiversity depend on relatively little-known differences in endemism among marine areas, although there is enough information to allow some to emerge (Norse, 1993). Areas of high marine endemism tend to be common in temperate and marginally tropical regions, where temperature gradients with latitude are steep, or where there is shelter from major environmental fluctuations, and also isolated islands or oceanic basins (Box 9.2). Based on this information, a global representative system of marine and coastal areas is possible. Kelleher & Bleakley (1992), for example, derived a list of 100–150 areas which they considered to be priorities for the conservation of global marine biodiversity.

Box 9.2 Areas of high marine endemism
(from Gage & Tyler, 1991; Norse, 1993)

- The northern (Senegalese) and southern (Angolan) limits of the West African maritime province;
- waters off south-eastern Brazil and nearby parts of Uruguay and Argentina;

- the northern (Sea of Cortéz, México) and southern (Ecuador and northern Perú) limits of the Panamic maritime province in the eastern Pacific;
- warm-temperate Japan and nearby waters off Korea and China;
- waters off the south-western Cape of South Africa;
- temperate Australian waters.
- the Okhotsk Sea and Kurile Islands in the north-west Pacific;
- the continental coast of northern South America;
- the South Atlantic oceanic islands (especially St Helena, Ascension, and Fernando de Noronha);
- the Red, Coral, Mediterranean, and Arctic Ocean-Norwegian Seas;
- the islands of Polynesia (especially Hawai'i, the Marquesas, Easter Island, the Societies, and Tuamotus);
- the Galápagos Islands; and
- the coastal waters of Antarctica.

9.3 Issues for individual species

9.3.1 Decline and extinction

The fossil record shows that all species eventually become extinct, and that the average lifespan of a species is about 4 million years (WCMC, 1992). Extinction rates vary greatly, however, and several incidents of mass extinction are known (Raup, 1988). One was in the late Permian from 250 to 245 million years ago, during a time of sudden change in sea level, volcanic activity, and climate. Another was at the end of the Cretaceous, about 66 million years ago, when the dinosaurs, plesiosaurs, and pterosaurs died out, probably due to sudden climate change caused by a cometary impact. At times of rapid environmental change, therefore, some species may be better adapted to new circumstances than others, for example by having greater behavioural flexibility. Some lineages may also be able to evolve fast enough to colonize new habitats in time and space, while others are unable to do so. The world is currently in another period of rapid change, this time caused by people, and many species are threatened with extinction as a result (e.g. Pimm, 1995). Like those in the Permian and Cretaceous, however, modern species also vary in their vulnerability to extinction (Box 9.3).

Box 9.3 Ecological factors which may make species more vulnerable to extinction (adapted from WCMC, 1992)

Rarity: some species are naturally rarer than others, and fragments of their natural habitat will contain fewer individuals; these small populations are at risk from random dangers such as genetic problems and disease.

Dispersal ability: species which cannot move easily among habitat fragments, or from large refuges to distant fragments, will tend to comprise small breeding populations each of which will be vulnerable.

Specialization: species are vulnerable if they are adapted to exploit rare or patchy resources, or if they depend on other (keystone) species which may become extinct.

Niche location: species which avoid the edges of natural habitats will be at a disadvantage after habitat fragmentation, when the proportion of edge habitat increases greatly.

Variable population size: species with populations that tend to fluctuate greatly in numbers over time may, after disturbance, overshoot the critical minimum number needed for recovery in the new or modified habitat.

Trophic status: species at higher trophic levels, such as carnivores relative to herbivores, tend to have smaller, less robust populations.

Adult mortality rate: populations of species with high adult mortality rates require constant replenishment, and this may be interrupted by disturbance, causing extinction.

Longevity: short-lived species may be more vulnerable than long-lived ones, since individuals cannot outlive episodes of disturbance and reproduce again afterwards.

Intrinsic rate of population increase: species whose populations can expand only slowly tend to recover poorly from temporary difficulties.

Although the extinction-proneness of populations which are already small has received much attention from conservation biologists, attention is now increasingly being focussed on the factors which cause populations to decline in the first place (Caughley, 1994). These are likely to be closely related to the problems which practical conservation action seeks to overcome. This approach implies that if it is to be effective, action should be designed only after the causes of decline are clearly understood for each species or population. This applies particularly where individual species are the focus of concern, but the large numbers of species in natural communities means that their collective response to disturbance is complex and hard to predict.

Many species which live in tropical habitats have very specialized ecological needs, and may become locally extinct if their habitat is disturbed even slightly. This does not mean, however, that disturbed tropical habitats are always of little value for conservation. A widespread but rare and specialized species may well survive a single episode of moderate disturbance, such as selective logging in tropical forest, since a large area of logged forest will contain intact refuges from which recolonization can occur. The danger to biodiversity from logging is rather that it is combined with fragmentation of natural habitats by settlements and plantations, or that it is repeated, or that it allows the entry of more destructive agents such as fire. These are all good reasons to prevent logging in nature reserves, but not necessarily to exclude logged forest from those reserves.

Many species can survive in modified natural habitats, and become extinct only when those habitats are converted to artificial ones. Environments derived from the natural state often include fragments of natural habitat set in a matrix of artificial habitats. Many native species will survive for a while in such a patchwork environment, especially those which are able to persist in the matrix itself (Laurance, 1991). Especially in the tropics a large proportion will become extinct, however, as the community adjusts to the dominance of small habitat islands and edges, and as people continue to disturb the area's natural habitats. The condition of the patchwork surrounding or between nature reserves will influence the effective population sizes of those native species which can move through it or live in it.

This means that a project should seek to manage patchworks so as to maintain as much natural habitat as possible, including connected strips along which native species can disperse if the strips or corridors are broad enough (Bonner, 1994). An important aim is thus to increase effective population size, in the hope that this will reduce the likelihood of local species extinctions. Because of unexpected events in nature, however, it is not possible to guarantee the long-term persistence even of large populations (Mangel & Tier, 1994). This implies that duplicate populations in separate nature reserves are desirable, and that contingency plans should be made to help managers minimize the impact of catastrophes affecting any one reserve.

A few species are able to take advantage of favourable conditions in artificial habitats, and thrive there. This is because new conditions may imitate some rare or transient feature of the natural system to which they are adapted. An early successional or highly mobile species which depends on colonizing tree falls, burned areas, or land slips in natural forest would be expected to take advantage of widespread disturbance. These are typically the weed species which live in association with people and farmland, and which easily colonize new habitats, such as islands, when local disturbance creates an opportunity to do so.

The fate of individual species in the wild is closely linked to the condition of their habitats, since it is very hard to exterminate a species within a large area of natural habitat, but easy to do so if the environment is badly disturbed (Caughley, 1977). There are exceptions, however, especially involving species which have low intrinsic rates of increase and are commercially valuable, making them vulnerable to overharvesting. These include rhinos (Bradley

Martin, 1983) and gaharu trees (*Aetoxylon*, Chapter 2). Assisting such species depends on suppressing exploitation within refuges, and suppressing harvesting and trade elsewhere. This may involve preparing species survival plans, and putting them into effect through legislation, supervision, enforcement, awareness-raising, and sometimes also the use of techniques for captive propagation, rehabilitation, and translocation (e.g. Riney, 1982; MacKinnon *et al.*, 1986; Nielsen & Brown, 1988; Olney, Mace, & Feistner, 1993; Bowles & Whelan, 1994).

Some species in a nature reserve may attack crops on nearby farms, and may be hunted as a result. Large mammals (e.g. elephants) can cause substantial economic losses while others (e.g. monkeys) may cause less real damage but are regarded as a nuisance and are specially persecuted because of this. The attitude of local people to a wild species will depend on their perception of its value to them relative to the threat it may pose to them. In Zimbabwe, for example, the CAMPFIRE programme changed attitudes to large ungulates by helping rural people to own them and to benefit financially from harvesting them (Pye-Smith & Feyerabend, 1994). As a result, certain villages became willing to provide fodder for hippos and dig wells for elephants during severe droughts, despite a long history of conflict with these species as crop raiders.

9.3.2 *Species in captivity*

Most species and genetic resources, if they are to survive at all, will do so in the wild, within viable, self-supporting natural ecosystems. The protection of species within such ecosystems is called on-site (or *in-situ*) conservation work. This distinguishes such efforts from off-site (or *ex-situ*) work in facilities such as zoos, botanical gardens and germplasm collections. On-site conservation is almost always a more efficient use of resources than off-site, because of the larger number of species involved and the lower long-term maintenance cost per species saved. If offered the choice, therefore, a rational conservationist would often prefer to let some species go extinct, than to divert scarce funds away from habitat protection. At certain times, however, and in the case of certain species and lineages, the choice might go the other way, and off-site methods should be viewed as complementary to on-site ones.

There are a several reasons other than protecting biodiversity why off-site conservation might be attempted, including the need to keep wild species or samples of them for educational or research reasons. In such cases, however, they should not be paid for out of conservation budgets. The guiding principle for project designers should be to ensure that the two approaches complement each other rather than compete for resources. Where funds are available specifically for conserving biodiversity off site, they should be reserved for the captive propagation of selected, endangered, or valuable species where this can be achieved at reasonable cost. In other cases, it should be remembered that different kinds of equipment and skills, and different levels of investment will be needed, according to whether the aim is to maintain animal or plant germplasm, whole plants, or invertebrate or vertebrate animals.

According to *The Botanic Gardens Conservation Strategy* (WWF & IUCN, 1989), priorities for conserving wild plant species in botanic gardens include maintaining rare and endangered species, and species or lineages which are economically important or have other uses such as ornamentation. Other priorities are to save species which are needed for restoring ecosystems, keystone species whose loss would cause other extinctions, and taxonomically isolated species with a high degree of uniqueness and scientific interest. Where a botanical garden is associated with an isolated and endangered ecosystem, its main role should be to save local species which are expected to become extinct in the wild. The same principle would apply where botanical gardens in one country are to propagate endangered species from another, as for example in a proposed 'emergency rescue operation' for threatened Philippine plants by a consortium of botanical gardens in Hawai'i, USA (Kristiansen *et al.*, 1993).

This implies the need for field inventories to establish which species in an area are most at risk, and may be considered as candidates for captive propagation. Species of conservation concern in an isolated ecosystem, however, may include all those with small population sizes, which even in a small tropical site may include thousands of plant and invertebrate species. This implies that off-site work should be carefully targetted, and may also need to include local species of the fauna as well as the flora. A balanced approach should be adapted to local circumstances, and should involve promoting community education, ecological research, and training. It should also involve collaboration with other botanic gardens and conservation groups, and the use of ecotourism as a source of revenues to support conservation activities.

Propagating animals in captivity poses a number of problems. Few proven techniques yet exist for working with invertebrates (Samways, 1994), although some butterflies are an exception (Collins & Morris, 1985). Captive maintenance of vertebrates, and especially large birds and mammals, is problematical for other reasons. Animals from such collections can only rarely be reintroduced to the wild (Kavanagh & Caldecott, 1988), and maintaining them is very expensive compared to other conservation investments (Balmford, Leader-Williams, & Green, in press; Balmford, Mace, & Leader-Williams, in press). Education is probably the single most important function of modern zoos, but a badly maintained zoo will send the wrong messages to the public, while still diverting money, attention, and political leadership away from more important conservation issues.

Some zoos are effective in educating the public and decision makers about conservation issues, and can also collaborate to maintain vertebrate species in viable breeding populations (Primack, 1993). These can be used to restock habitat areas where those species have been lost for reasons that no longer apply (see below). Zoos can also generate funds in support of field conservation work, especially where this helps the species and ecosystems on display at the zoo, and they can support captive research which may help some conservation projects. In making decisions about maintaining animals in captivity, an important guide should be the likely maintenance cost of each species relative to the feasibility of reintroducing it to the wild, and its educational value and attractiveness to visitors. According to these criteria, reptiles, amphibians, and invertebrates will often be more feasible candidates for off-site biodiversity management than are large mammals and birds.

The issue of captive wild mammals often arises in project areas, since hunting cultures tend to create a supply of young animals, and they may be given to project staff or confiscated by the wildlife authorities. This calls for a project policy on keeping wildlife as pets, and also on what to do with animals that have been taken illegally from the wild but cannot be returned there. Such animals demand veterinary and other care, and absorb resources which might otherwise be spent on protecting habitats or suppressing hunting. The normal policy of a project towards confiscated and captive animals should therefore be to kill them humanely.

Exceptions might be made for great apes, which many people believe belong with humans in a moral 'community of equals' qualifying them for special care (Cavalieri & Singer, 1993), or for animals of such rare species that it is reasonable to attempt captive breeding. An example is provided by the drill (*Mandrillus leucophaeus*), an endangered West African monkey threatened mainly by hunting (Lee, Thornback, & Bennett, 1988; Gadsby, 1990; Oates *et al.*, 1990; Olney, Mace & Feistner, 1993). The species occurs in both Divisions of Cross River National Park in Nigeria (Chapter 3) and in the nearby Korup National Park in Cameroon, but extinction seemed possible prior to full development of these protected areas. A project was therefore started in 1990 to gather confiscated drills at a captive breeding facility in Calabar. By early 1995, this contained 27 drills in the world's largest captive group, and three births had occurred (Gadsby & Jenkins, 1995).

The drill provides an example of the use of the 'declining-population paradigm' (as opposed to the 'small-population paradigm') in deciding how to solve conservation problems affecting individual species. As described by Caughley (1994), this would ideally involve at least five steps:

Step 1: deduce why the population declined and which agent caused the decline.

Step 2: remove or neutralize the agent of decline.

Step 3: confirm that the agent of decline has been correctly identified (e.g. by releasing and observing a 'probe group').

Step 4: if results are promising, restore the wild population by translocation or captive breeding (as close to the site as possible) and release.

Step 5: monitor the re-establishment of the population.

In the case of the drill, the decision to establish a near-site facility to rehabilitate (i.e. rescue and care for) captive monkeys, and to breed them in captivity, was made after exhaustive field studies which showed hunting to be the primary agent of decline. The decision took into account the fact that large areas of habitat would be protected within national parks, that hunting in those areas would eventually be greatly reduced, and that there would thus be suitable sites for release of salvaged individuals. Caughley's other steps remained to be completed for the drill in early 1995, and it was envisioned that the whole cycle of habitat protection, hunting suppression, and drill re-establishment would take several more years.

Caughley (1994) describes one of the few cases in which all steps were undertaken correctly. This involved the Lord Howe woodhen (*Tricholimnas sylvestris*), a flightless rail which is endemic to a small, isolated island in the south-west Pacific. Lord Howe Island was not settled by people until the nineteenth century, and its rail is thought to represent some hundreds of related species which were endemic to other Pacific islands, but which became extinct due to human contact over the last two millennia. By the late 1960s, the Lord Howe rail was reduced to fewer than ten breeding pairs living in montane habitat. Long-term field studies showed that the reason for this was that their refuge was inaccessible to pigs, and ruled out other factors such as the distribution of food supply, lack of suitable habitat, and predation by rats. This allowed conservation action to be focussed on eliminating pigs, which were duly exterminated on the island. It should be noted in passing that action against exotic animals such as pigs and goats would not normally need special justification, since they often help to cause extinctions on small islands.

The Lord Howe rail programme continued with near-site captive breeding to provide additional rails for release elsewhere on the island. Probe groups were released in 1981–2 and closely monitored until they began breeding, whereupon additional captive progeny were released and the breeding programme was closed in 1983. Within ten years, there were up to 60 breeding pairs in the wild. As Caughley (1994: 237) notes: 'The steps followed for the woodhen – diagnose the agent of decline, neutralize the agent of decline, re-establish the species of concern – may serve for almost any other troubled species.' These steps are appealingly obvious, but there are many other cases where failing to follow them all has resulted in efforts which are ineffective and needlessly expensive (Caughley, 1994 cites the example of the California condor, *Gymnogyps californianus*).

9.4 Managing nature reserves

9.4.1 Management plans and paper parks

Managing a nature reserve means making sure that it is used only in ways that help or at least do not hinder the survival of its species and ecosystems in the long term. This involves setting and following guidelines for using the area in ways that help people perceive its value as a reserve. These guidelines will often be specific to zones within the reserve, and a management plan involves defining these zones and guidelines. Another important role of such a plan is to make sure that a reserve's boundaries are in the right place, since key areas may have been left out when the area was legally constituted, or conditions may have changed in parts of the area since then. The plan will then define what needs to be done, and assign responsibilities within the management service. It will define what are acceptable patterns of visitor use, and design programmes to monitor environmental changes and the impact of management measures. It will indicate measures to increase public support for the reserve, and design in-service training arrangements for employees of the management authority.

It is important to bear in mind the series of distinct steps leading to an area being genuinely protected:

- first, it will be identified as meriting protection;
- then it will be legally proclaimed (often by publishing a description in an official gazette);
- then a management plan will be prepared;
- then the management plan will be implemented; and

• finally, as a result, a new and stable equilibrium will be achieved in and around the area.

Many nature reserves never make it beyond the planning stage, at which point they are still 'paper parks'. There are many reasons for this, one of which is that identifying areas and writing management plans are often undertaken as office jobs, or else with only short-term field excursions from an office base. Putting a management plan into effect, by contrast, demands long-term commitment of time in the field, with huge consequences for the careers of the personnel involved and the agency which employs them (Diamond, 1986; Janzen, 1992a).

Another reason for the persistence of paper parks is that management plans often call for at least 5–10 years of work. This work needs to be guided, managed, monitored, funded, and supported logistically, often in a remote area with poor infrastructure and communications. Moreover, a commitment to implement a plan means guaranteeing a budget far into the future. Even if annual costs are kept low (e.g. by using local staff and working closely with local people), the total cost will add up over the years to a size which will deter many governments and donor agencies.

Management plans should relate to current conditions if they are to be useful, and they can quickly become out of date. Conservation agencies thus collect obsolete management plans even faster than they do paper parks. The ideal is for planning and implementation to be linked within one continuous process, but this seldom happens (MacKinnon, 1994b). A project should always be designed to strengthen links between plans and actions, the former being matched both to the threats which they seek to address and to the resources which are available at known times. This will help to avoid confusion between plans (i.e. what will be done with existing resources) and proposals (i.e. what would be done with new or additional resources, should they be obtained). This confusion can arise when a single project document contains both plans and proposals, either because promised resources may not turn up on time, or because it is hoped that the act of planning may cause extra resources to be obtained.

A reserve will remain a paper park until management occurs, and in view of the short currency of management plans there is little point in planning unless resources are available to be spent on managing (Box 9.4). The fact that a reserve is only a paper park may be no cause for concern if it is not threatened, but this status is not adequate if threats exist. This calls for an informed judgement on the kind of threat which may be involved, before investment is made in planning. Such judgements will be affected by the total resources available for conservation in the country concerned. If resources are abundant, then lower levels of threat may justify management planning, although at very low levels of threat it would be hard to justify more than a plan for research and monitoring activity.

Box 9.4 Guidelines for effective conservation planning
(adapted from MacKinnon, 1994b)

Keep plans clear and brief. Sections describing what needs to be done should be short and presented in clear points, as a series of confident, decisive instructions in a style which can be welcomed and used by a busy reader.

Present documents in the local language. Even if leaders are fluent in an international language, all plans should also be presented in the working language of the country or region concerned. To make plans more broadly accessible, extra translation effort is needed in countries (such as the Philippines) where many citizens are not fluent in the official language.

Plans should build on each other. They should also not compete or conflict with other or pre-existing plans (except where previous recommendations are seriously flawed or obsolete).

Plans should be developed with agencies capable of implementing them. Many plans are commissioned by agencies other than those with a mandate to pay for or manage their implementation. This can lead to a mismatch among institutional priorities, under-motivated counterparts, and other problems which prevent action.

Keep plans realistic. They should not be overloaded and turned into dream plans. Are there really the people to do the job? Will there be enough funds? If in doubt, plan what must be done with assured resources, and then add a separate section of additional needs if further resources can be obtained (see text).

Adapt plans to precise needs, and omit irrelevant ideas. A plan to acquire land for conservation should be left as a land-use or spatial plan; one to obtain an official budget allocation should be tightly focussed on the justification for investment; while one for use by a donor agency should be tailored to meet its specific objectives and procedures.

Another factor is that planning work in a reserve can at times be counter-productive. It might waste resources which would be better spent elsewhere, and it might create new threats to the reserve. This may occur when access roads and buildings are made to support the planning team, if they attract visitors or settlers later. Planning to manage a reserve also implies a claim to regulate other people's use of it. This may provoke action by other claimants, ranging from logging or mining companies to local people who may already use the area. The former might have seen the reserve as a future resource asset, while the latter may see it as their traditional property. Demarcating and zoning a reserve, even on paper, may thus cause it to become more threatened, or else may provoke an unnecessary public-relations crisis, or both. Where such claims might arise, other judgements are needed on whether there are ways to control competing demands on the resource, or whether local people pose a real threat to the reserve. If the answer is 'no' in either case, it may be better to leave the area alone. If the answer is 'yes', then a more vigorous response than planning alone will be needed.

9.4.2 Reserve management by zone

If a nature reserve is seriously threatened, then its management plan must specify sufficient action to neutralize those threats. If the reserve is zoned, then different kinds of effort can be distributed among the zones as needed, and progress and impacts can be monitored and evaluated against the aims of each zone (Box 9.5). The most important parts of a reserve for biodiversity conservation are those whose ecosystems best represent local natural conditions, and which are most intact and most strongly protected from disturbance. These can be called **core areas** or **biodiversity refuges**. Such zones are normally reserved for very low-impact uses such as monitoring, research, and controlled use by visitors. Once conservation priorities and demand for access have been determined, a core area might be subdivided into **exclusion** or **research zones** (not for public access), **wilderness zones** (for low-impact tourism), and **limited use zones** (for certain other kinds of access). To meet the needs of visitors, parts of the core area may have to be reclassified as **intensive use zones**, where trails, campsites and viewing points can be located.

Other zones go under various names internationally, but there are three main options. Firstly, there are **recuperation zones**, which are those areas previously disturbed by farming, logging, or intensive hunting, fishing, and gathering, but which have been absorbed by the nature reserve and are being allowed to regenerate free of such disturbance. Secondly, there are **infrastructure zones**, which are those areas needed for buildings, roads, or other intensive uses in support of administration, maintenance, public services, and control programmes. Lastly, there are **traditional use zones**, where permission may be granted to certain local people to use resources, usually in exchange for their assistance in managing the reserve as a whole (see below).

Box 9.5 Kinds of management zone within nature reserves

Core areas or biodiversity refuges: intact natural areas set aside for protecting and studying biodiversity, monitoring the environment, and for low-impact use by visitors, including:

- *exclusion zones* (no access at all; aerial or boundary surveillance only);
- *research zones* (access by research and protection staff only);
- *wilderness zones* (for low-impact tourism); and
- *limited use zones* (for other limited recreational access).

Intensive use zones: natural areas set aside to accommodate dense trail networks, camp-sites and viewing points to service public needs.

Recuperation zones: disturbed areas set aside for recovery from agriculture, logging or intensive hunting and gathering.

Infrastructure zones: disturbed areas set aside for conversion to such artificial habitats as roads, buildings, carparks, and waste-disposal sites.

Traditional use zones: natural habitats set aside for the continued use of certain resources by specified communities, in exchange for their assistance in managing the nature reserve as a whole.

9.4.3 Traditional use zones

The rationale for making a traditional use zone is as follows. Most natural environments have long been harvested by local people, who take meat, fish, plant parts, or other products from the area. This harvesting may or may not have had a detectable impact on species populations or other features of the ecosystems which a nature reserve seeks to protect. In either case, excluding people from the area will cause its former users to suffer at least some economic loss, and may also offend them and cause hostility to the reserve. These represent costs to the project, because of the need to pay compensation in some form, and/or to invest in public relations or enforcement activities.

A project's designers may or may not decide to accept these costs. This judgement will depend on factors such as the resources available, the vulnerability of certain species to continued harvesting, the likelihood that harvesting patterns will become more damaging in future, and the strength of the case which current users can make for being allowed to carry on harvesting. The latter may include a rejection of the right of any outsider to impose restrictions (see Peluso, 1993). If for whatever reason the costs of excluding local harvesters are too high, a traditional use zone may have a role in a compromise arrangement. This will often involve current users agreeing to be bound by new controls, in exchange for official recognition of their exclusive right to harvest resources within the zone. This transaction should give local people an incentive to help protect the reserve's boundaries, to use its resources with care, and to co-operate with the management régime in other ways.

Common threats to nature reserves include road building, industrial exploitation of timber, fish or minerals, and the development of commercial plantations. Relative to these, it is often fairly assumed that continued harvesting by small numbers of people using traditional technologies is not a serious problem. This assumption should be reviewed in each case, using information obtained during the project design process. This is because there are many cases in which traditional hunting is far from being a benign factor (e.g. Redford, 1989, 1991). A formal test which might be applied from the point of view of biodiversity conservation is provided by IUCN's (1993, 1994) definition of sustainable use of wild species.

According to this, there should be no reduction in the potential of a harvested species to be used in the future, and also no reduction in the long term viability of that species, any other species, or the ecosystems which support and depend upon them. When applied to a species, the idea of maintaining viability means that no harm is done to the capacity of the species to maintain its genetic diversity, or to its potential for evolution, or to its likelihood of survival. When applied to an ecosystem, it means that no harm is done to the capacity of the ecosystem to maintain the diversity of its habitats, species, and genetic lineages, or to its capacity for continuity and renewal, or to its productivity.

This is a rigorous definition, and may defy application in real-life circumstances. As observed by MacKinnon (1994b): 'One can try to recommend sustainable utilisation in some areas or at some times but given the fact that sustainable use is itself an almost impossible concept to achieve, it is vital to also include a large chunk of old-fashioned protection effort into any programme that wants to claim to be conserving biodiversity. It is much easier to stipulate what one may or may not harvest than to try to calculate a sustainable level of harvest. Closed-seasons, off-limits, total protection items, certain numbers to be returned to the wild or spared in the harvest (size limits) etc., are ways that numbers of surviving individuals can be regulated.'

The use of traditional use zones should be monitored, with the aim of ensuring that only permitted users are present, and that neither legal nor illegal harvesting is threatening any species. Over-harvesting of any species may be cause for further negotiation, guidance, or enforcement of restrictions depending on circumstances. If traditional harvesting is considered to be benign and there are no other threats, many reserves might just as well be left wholly to traditional users. In such cases, there may be complete overlap between the concept of a nature reserve with that of a communal forest or indigenous people's reserve.

9.5 Knowing what is being saved

Conservationists believe that the components of biodiversity should be saved and studied at the same time, that studies will show how to use them sustainably and how to teach about them interestingly, and that both use and interest will help people support the idea of saving them. Wild species differ in their vulnerability to different threats, so their particular nature must be known if they are to be saved. They also differ in how they might be used safely and productively, and conservation projects need to find ways for local species to support local farming and harvesting systems, to provide ecological services and attract tourists, or to reward the efforts of biodiversity prospectors.

The starting-point for all such work is to make an inventory of the components of biodiversity in a reserve, which has two main aims. The first is to identify those species most at risk of extinction in the reserve, which will allow protection efforts to be focussed effectively. These may include off-site captive propagation and later re-establishment in the wild, provided the causes of decline are known and can be neutralized (see above). The other main role of an inventory is to start the process of gathering and managing data for potential commercial and other use, for example through biodiversity prospecting (Sittenfeld & Gámez, 1993).

Faunal and floral inventories of nature reserves were once little more than species lists and maps of the main ecosystem types, but information technology has transformed this concept. A biodiversity inventory should now relate to a database which has been designed to allow data to be managed with specific aims in mind. A modern database should be able to meet the needs of a variety of uses and users, ranging from local people, teachers, and tourists to prospectors, planners, and scientists. It should also be able to relate local information to several kinds of national data, such as digitized maps and locality records from national collections of species. Such a database could be used, for example, as a tool for spatial planning and environmental impact management.

A reserve's inventory should aim to identify all species to high taxonomic standards, to document them in enough detail that specimens can be located in the future, and to gather additional data on their ecology and ethnobiology.

It may also be designed to allow certain kinds of data to be offered in a marketable form, and to permit the sale of those data for the benefit of local people and to help to pay for managing the reserve. The inventory process should also be used to teach local people about biodiversity. Before undertaking a local biodiversity inventory, however, the questions in Box 9.6 should have been asked and answered in detail.

Box 9.6 Questions to ask before undertaking a biodiversity inventory for a nature reserve (adapted from Janzen, 1992b; see also Janzen, Hallwachs, Jiminez, & Gámez, 1993)

Is a complete inventory of the area feasible? This should be considered in relation to contemporary techniques of personnel management and taxonomy; practical difficulties would need to be analyzed in the local context and solutions determined.

Can it be done without damaging the area's biodiversity? This should lead to guidelines and operational techniques to minimize damage.

Is the reserve an appropriate site for an intense inventory? This will focus attention on the biological, logistical, political, sociological and economic attributes of the area.

What level of inventory can be achieved? Most major taxa can be inventoried adequately and quickly using current techniques if efforts are focussed and managed properly. Problems and exceptions need to be identified and potential solutions to problems targeted for study.

What protocols would be used in the inventory? There is a need to determine the general administrative, technological, and sociological protocols, the structure of the inventory teams, involvement of the international taxonomic community, and integration of user needs with data collection and management procedures.

What would be the products of the inventory? These could include taxonomic, locational, ecological, and ethnobiological information on the organisms present in the site, in each case opening a permanent 'file' on the organism concerned for later elaboration through further research.

Who would use the products of the inventory? Expected users would include research scientists, tourists, taxonomists, educators, reserve managers, biodiversity prospectors, legislators, politicians, and resource planners. These and other users should be identified and their needs assessed.

How much will the inventory cost? Budget estimates would have to be devised on the basis of a detailed plan of work. Significant capital investment will be needed in facilities to manage specimens and data in the project area.

The data management needs of conservation projects and agencies will vary, but they will usually include associating physical specimens with names, descriptions, and records of natural history. No international consensus had emerged by 1994 on how best to collect and manage biodiversity data, but the leading information standards and protocols appeared to be the HISPID standards (Croft, 1992), which are specifically botany-oriented, or those of the Association of Systematics Collections, which merge experience from the management both of botanical and of entomological collections (ASC, 1993).

Key guidelines for designing a biodiversity database include the following (A. Allison, personal communication, March 1993):

Guideline 1: To allow for taxonomic uncertainty, records of individual specimens should be kept separate from one another and not pooled permanently as 'species' or 'occurrences'. Thus species names should be separated from genus names, rather than stored in one field.

Guideline 2: Temporary pooling of specimen-based records can be used to call up literature or to superimpose on maps from elsewhere in the database, but software should be used which will ensure that taxonomic changes at the specimen level are reflected automatically in such pooling.

Guideline 3: Graphical records, such as pictures and diagrams associated with species, should be kept in a separate graphics unit of the database.

Guideline 4: The sites of origin of all specimens should be located within a geographical information system (GIS), which should be layered, and soils, topographic, vegetation and other kinds of information should be kept separate but available to be called up in various combinations to answer different kinds of questions.

An advanced system such as that developed by the Intergraph Corporation and INBio in Costa Rica (Chapter 6) seeks to keep very large amounts of alphanumeric (specimens, species and literature), graphical, and GIS data separate but interactive. Not all biodiversity database users will have the same needs, however, so the design in each case should be guided by the objectives. This is probably the most important point about database design, that the needs of the users of biodiversity data in each case should be understood before any investment is made. Data may be collected meanwhile, but all information should be kept separate so that it can be manipulated later.

10
Options for development

10.1 Introduction

The main task in managing a nature reserve is to ensure that people intrude on the reserve only as staff, visitors, and authorized harvesters, and that little harm is done when they do so. More serious threats to reserves come from outside, in the form of competing demands on resources both planned and unplanned. These include officially approved industrial exploitation, conversion or infrastructure development within the reserve, as well as informal encroachment, colonization, and open-access harvesting there. To relieve them, a project must extend its area of interest beyond reserve boundaries, and seek to neutralize the causes of threats where they originate.

Those parts of a project area which are outside a reserve thus give rise to the most complex social, economic and political problems. These are often the most difficult and expensive to solve, but when solved they contribute most to the long-term security of the reserve. For design purposes, a project area outside a reserve can be divided into different zones, which will allow zone-specific issues to be identified and their solutions planned and budgeted. One important kind is that of the buffer zone, but others occur under various names according to their purpose. Because all these zones are outside the nature reserve, they cannot usually be controlled by reserve managers. Instead, their management will depend on agreements being negotiated with and implemented by local people and other interest groups.

10.2 Zone options

10.2.1 Buffer zones

The concept of a **buffer zone** is that of an area of mainly natural habitat, outside but adjacent to a nature reserve, where the use of resources by people

201

is monitored or modified in ways which help support the reserve (Cartwright & Frame, 1986; MacKinnon *et al.*, 1986). In tropical forest, this might involve the careful planning and supervision of logging in areas close to a reserve; an example being that of forestry buffer zones around Korup National Park in Cameroon (Tchounkoue *et al.*, 1990). The aim would be to create a region of modified natural forest, in which the degree of disturbance decreases as the reserve is approached. The reserve and buffer zone would therefore continue to share populations of many native species, resulting in better species survival prospects in the area as a whole. With the reserve boundary located within regenerating forest, ecological edge effects should also be reduced, which may otherwise intrude up to several kilometres into the park's ecosystems (Janzen, 1986b).

In rare cases, a buffer zone might be created to meet the needs of only one species. An example was the proposal to help protect rhinos in Sarawak's Pulong Tau National Park (Chapter 2). In this plan, forestry coupes (annual felling zones) outside the park were to be logged, but only in a sequence which would allow rhinos to move away from the logging in the direction of the park upstream. Meanwhile, the Forest Department was to prevent poaching by controlling access to the buffer zone, and by closely supervising the work of timber workers throughout the area.

The idea of a buffer zone makes most intuitive sense to conservationists when it implies insulating intact natural ecosystems from artificial ones. From this point of view, a buffer zone is primarily an adjunct to a nearby nature reserve. To people living around a reserve, however, a buffer zone can be important not because it helps protect the reserve from harm, but because it helps protect local people from hardship. Other groups may have a different viewpoint, and may see the same area, for example, as a source of raw materials for industrial exploitation. Conflicts of interest among these groups can be hard to resolve, since buffer zones are outside reserves and are therefore often administered not by the government agency responsible for conservation, but by various others. With several agencies involved, there is the need to negotiate an early agreement on what the buffer zone is for, and to establish a forum where people can talk through any problems that arise later.

10.2.2 Other zones

Communities that are close to a nature reserve may be thought of as living in a **support zone**. This is because their needs, and the actions they take in meeting them, will have a special role as the project unfolds. They might become either dangerous antagonists or strong supporters of the reserve, and action can be taken in the support zone to encourage them to help protect and develop the reserve. This action may include promoting sustainable forms of land use through training, negotiation, education, and subsidy. The support zone concept also has a role in public relations, since (unlike that of the buffer zone) it implies a mutually supportive relationship between the reserve and the local people, rather than a defensive posture against them.

Various kinds of **special zone** may also be included in a project area. These identify areas and populations outside the main focus of a project, but which are nevertheless of actual or potential significance to it. They may be used to create budgets for additional research or special habitat management action. **Education zones**, for example, may extend far from a reserve and help to define areas where public understanding of the project is to be promoted. **Fire control zones** focus attention on the special need in certain areas to prevent and suppress forest fires. **Investment zones** can be used to promote private-sector involvement in low-impact forestry, rattan-planting, and eco-tourism. This may involve using reciprocal taxation agreements, tax holidays, and pioneer status regulations to encourage particular investments. All these special zones should contribute to the project as a whole, but they may or may not overlap with one another in time or space. Their purpose is to highlight the need for resources with which to accomplish aims specific to each zone.

10.3 Managing harvests

10.3.1 Introduction

Natural ecosystems, and particularly tropical forests, can provide many different products for use and consumption by people (e.g. Stiles, 1994). If ecosystems are not too badly damaged by this harvesting, project designers can arrange for local people to benefit from it, for example within buffer zones and traditional use zones. One option is for a project to work with local people to establish which products are or could be harvested, and to help them work out a system for doing so sustainably. This should include arrangements for monitoring the impact of each harvest on others, and on the local ecosystem as a whole. The single most important rule for avoiding overharvesting is to prevent open-access, competitive exploitation of the same resource by more than one group of people. It is always better to base management strategies on exclusive access by known people, with whom a co-operative relationship can be developed over time.

10.3.2 Woods and canes

Many tree species are used as sources of wood, but these are overharvested wherever modern technology and large markets for wood coexist. Overharvesting can only be avoided by closely following certain rules, such as those given in Box 10.1. Apart from the need to retain the forest as forest, the most important and universal of these rules are to find out what is in the forest before it is logged, to respect an appropriate minimum girth for felling, and to monitor impacts on the forest (R. G. Lowe, personal communication, 1991). These rules are simple, but they require a social and political consensus, discipline, commitment, and a long-term perspective to be applied successfully.

Box 10.1   Rules for avoiding over-harvesting timber from natural forests (adapted from Collins, Sayer, & Whitmore, 1991)

Rule 1: the forest should be protected from unplanned use (e.g. theft of timber) or conversion (e.g. to farmland).

Rule 2: the annual cut must be realistically assessed.

Rule 3: annual felling zones should be clearly demarcated.

Rule 4: there should be detailed pre-felling inventories, to establish what is there before logging.

Rule 5: trees should be clearly marked to be retained or to be felled.

Rule 6: felling zones should be exploited only to acceptable and defined damage limits.

Rule 7: there should be a proper post-felling inventory, to establish what is left after logging.

Rule 8: the harvest should be checked by species to prevent only the most valuable timber from being selected.

Rule 9: the forest should be continuously inventoried and otherwise monitored, so that changes can be seen and understood.

Rule 10: main roads should be designed and maintained properly, and due attention given to controlling erosion on spur roads and skid trails.

A project may need to respond to local harvesting patterns which focus on particular woods with specialised uses and high commercial value. Examples include perfumed woods such as gaharu in South-east Asia (Chapter 2), sandalwood (*Santalum* spp.) in the Pacific (Applegate *et al.*, 1990), and cattle-herding and chewing sticks in south-eastern Nigeria (Chapter 3). These forms of harvesting can be destructive, although in some cases a sustainable supply can in principle be arranged: gaharu through off-site culture, and chewing stick through coppicing or pollarding (Okafor, 1989, 1990; Hurst & Thompson, 1994).

The commercial significance of the climbing palms or rattans was mentioned in Chapter 2. In Asia, where most species are native, several rattans of the genus *Calamus* can be harvested sustainably from managed forests, or deliberately planted there. Many rattans produce expanding clumps of stems by suckering, and are able to sustain regular harvesting of canes. Rattans also occur and are used by people in West and Central Africa. In southern Nigeria, for example, species of *Laccosperma* and *Eremospatha* are often used by rural people in making houses and household and farm goods (Okali, 1990).

As Asian rattan supplies have declined, African rattans have attracted increasing commercial interest, both for export and for local manufacture of furniture (Morakinyo, 1994). Canes are usually bought by dealers from local collectors, but low prices are paid to local people, and greater benefits could be obtained locally either by better marketing, or by developing local skills in cane processing before first sale. A project may be able to help a community devise a more rewarding and sustainable harvesting strategy. Since rattan canes are such versatile materials, and are best harvested from natural forests, they often have a useful place in buffer zone management around nature reserves in tropical forest areas.

10.3.3 Hunting and fishing

Terrestrial wildlife populations are harvested for meat, hide, horns, ivory, and many other products ranging from porcupine bezoar stones in Sarawak (Chapter 2) to deer-bone tonic wine in China (Chapter 5). Principles of wildlife management are well known, and sustainable harvests are feasible in any ecosystem provided the size or structure of populations are not too greatly distorted, and that their key ecological needs continue to be met (Caughley, 1977; Riney, 1982; MacKinnon *et al.*, 1986). These requirements imply the need for ecological and other research effort to be made before harvesting occurs, and for the effects of harvesting to be closely monitored.

A particular danger in many project areas is that a management strategy devised for one set of hunting techniques and market conditions can be upset by the arrival of new ones. Public access to shotgun ammunition increased suddenly in Sarawak in 1986, for example, because of a change in security rules by the police, and this caused increased pressure on wildlife populations (Caldecott, 1988a). Similarly, the introduction of shotguns and hunting dogs in Nigeria in the 1980s greatly increased pressure on drills (*Mandrillus leucophaeus*), mainly because this species tries to hide from dogs in trees where they can easily be shot (Gadsby, 1990). Wildlife management systems should thus have the flexibility to respond quickly to changes in hunting technology and trade patterns.

This also applies to fishing in project areas, since the difference between harvesting and overharvesting a fish population depends both on its productivity and on the technology used to catch fish. The underlying productivity of a fishery depends on the condition of the habitats of target species, particularly key feeding, spawning, and nursery areas, and on the integrity of migration routes. The impact of harvesting can change rapidly when traditional hooks, harpoons, traps, and vegetable poisons are replaced by more efficient nets, explosives, electricity, and industrial toxins.

Traditional poisoning, for example, was often a labour-intensive, social activity used rarely because of the labour costs involved in damming rivers, herding fish, and extracting the poison from vegetable matter. Traditional poisons also tended to be biodegradable and non-persistent, but commercial markets for fish encourage both more frequent use of poisons, and the use of modern toxins such as agricultural pesticides. As well as endangering the health of people, this can have disastrous effects on species-richness. A large amount of commercial insecticide is needed to disable fish, but very dilute residues can extinguish arthropod communities in long sections of river. This is the opposite of the effect of traditional *Derris* root poison, for example, which is much less toxic to arthropods than it is to fishes (Burkill, 1966).

As with hunting, the intensity of fishing can be greatly affected by transport and trading patterns, since previously unharvested fish in remote locations can suddenly be exposed to exploitation by urban markets. During the 1970s, for example, commercial speedboat routes penetrated the interior of Borneo. This allowed traders to invest in cold storage facilities to collect fish and game for sale in towns downstream (Caldecott, 1988a). These factors imply that monitoring and flexibility should be designed into all fishery management systems.

10.3.4 *Medicines*

People living in rural areas in the tropics often have little access to orthodox or modern medicines, and may depend on local plants for most or all of their medical needs. The capacity to do so depends on the ethnobiological information which their societies have accumulated over time (e.g. Thomas &

Tobias, 1987; Thomas *et al.*, 1989; Avé & Sunito, 1990; Leaman, Yusuf, & Sangat-Roemantyo, 1991; Plotkin & Famolare, 1992). Preparations are used to treat a wide range of ailments, including all kinds of infection, asthma, diabetes, and hypertension. Since they can be effective, access to native plants may have an important role in local health care, while also being culturally important and contributing to local feelings of self-sufficiency and confidence. Lastly, medicinal plants may be cultivated and surplus amounts sold outside the immediate area, thus creating sources of local income.

These preparations often have real effects on pathogens and symptoms. This may be because plants have evolved chemical defences against predation and disease which affect animal systems, or which inhibit fungal, bacterial and viral growth or reproduction (Janzen, 1975). Thus, many kinds of ortho-dox medicinal drugs originated in nature (Myers, 1984; Farnsworth, 1988). The following are well-known pharmaceuticals which were first discovered in tropical vegetation, and which might be mentioned when making a case for conserving biodiversity:

- reserpine, from an Indian shrub (*Rauwolfia serpentina*), which is used to reduce blood-pressure;
- tubocurarine, from a Brazilian vine (*Chondrodendron tomentosum*), which is used as a muscle relaxant in surgery;
- vincristine and vinblastine, from a Madagascan herb (*Catharanthus roseus*), which is used to treat certain cancers;
- diosgenin, from a genus of Mexican vines (*Dioscorea* spp.), which is used to prepare oral contraceptives;
- quinine-related anti-malarials, from a Peruvian tree (*Cinchona ledgeriana*);
- cocaine-related local anaesthetics, from an Andean shrub (*Erythroxylum coca*); and
- emetine, from a Brazilian shrub (*Cephaelis ipecacuanha*), which is used as an amoebicide.

Many other medicinal families may be discovered in future through biodiv-ersity prospecting, often guided by ethnobiology. This is a useful argument in favour of protecting biological and cultural diversity (Chapter 11), but, where ethnobiological knowledge is used to help prospecting, local people should benefit in proportion to the revenues obtained from any product based on their experience. It should also be considered how using these biodiversity assets can contribute directly to solving problems which threaten their con-tinued existence.

10.3.5 Foods from wild plants

Plants growing in and around reserves and buffer zones can meet many of the dietary needs of local people, whether continuously, seasonally, or in times of crop failure or other emergency (e.g. Purseglove, 1968, 1972; Jong, Stone, & Soepadmo, 1973; Okafor, 1979). These food items can include the fruits, nuts, and seeds of many species, as well as leafy vegetables, condiments, spices, oils, starches, and drinks such as palm wine. They may also be culturally important, as in the case of the cola nut (*Cola acuminata*) in West Africa and betel nut (*Areca catechu*) in South-east Asia, both of which are used in social ceremonies as well as being chewed privately as a mild stimulant. Some may also be sold outside the project area, for example bush mangos (*Irvingia gabonensis*) in Nigeria, which are widely traded because the powdered kernels give a mucous-like consistency to soups.

In some cases, valuable foodstuffs are obtained through the interaction of several species. This strengthens the case for an ecological approach to buffer zone management, rather than trying to simplify ecosystems to favour one particular kind of production. Examples include the culture of rhinoceros beetle larvae in the trunks of standing sago palms (*Metroxylon sagu*) in Siberut (Whitten, 1982), and the interdependency of certain orchids, euglossine bees, and Brazil nut trees (*Bertholletia excelsa*) in Amazonia (Lewington, 1991).

10.3.6 Agroforestry

The choice of farming system around a nature reserve can have an important influence on the reserve's future. Where native species are included, this may allow farmland to be used by a few species within the reserve, thereby extending the range of their populations. A far more critical question, though, is whether or not the farming system is productive enough to meet the needs of local people, and durable enough for this support to last a long time. If not, then the reserve itself may be steadily encroached upon until it is ultimately destroyed. An answer often lies in finding suitable agroforestry systems which can be established on farmlands within the project area.

Agroforestry is a general term for all farming systems which mix trees and woody shrubs with other crop plants and/or with animal husbandry. Because of the difficulty of maintaining soil fertility in high-rainfall areas without chemical inputs, the most durable agroforestry systems include nitrogen-fixing leguminous plants, often among the woody component of the vegetation. The aims of agroforestry are:

- to ensure that soils are protected from wind and water erosion, especially by covering them with vegetation and binding them with living root systems;
- to ensure that nutrients are replenished, especially by fixing nitrogen, by mulching with leafy wastes and by adding manure; and
- to ensure that useful production is as continuous and valuable as possible, especially by adding a range of economic trees which do not compete with one another or with crop plants, and which produce harvests at different times.

Durable agroforestry systems can be designed for almost any given combination of climate, soil type, and terrain, by putting together species of woody plants, non-woody plants, and livestock with known attributes (e.g. Kang, Wilson, & Lawson, 1984; Kang, Van der Kruis, & Couper, 1986). This technical approach can be made far more effective by using existing local knowledge of the uses and characteristics of wild and cultivated species, and local awareness of how to tend and market them (e.g. Chambers, 1994a, 1994b, 1994c). This information is an important resource for use in improving farming systems in a project area (e.g. Lowe, 1986; Abel *et al.*, 1989a, 1989b; Holland *et al.*, 1989; Okafor & Fernandes, 1987; Okafor, 1978, 1989, 1990; Okafor & Caldecott, 1990).

Project designers should also be aware that new agroforestry systems carry the potential risk of being too attractive as replacements for natural ecosystems. Agroforestry can be achieved even on very steep slopes, for example, if combined with terracing and barrier hedges of vetiver grass (*Vetiveria zizanioides*; NRC, 1993; World Bank, 1993). Once a tropical ecosystem is simplified by any kind of agriculture, the great majority of its original species richness will have been lost, even though it may seem verdant and diverse by temperate standards (Janzen, 1994). This is an important difference between tropical and non-tropical environments, since in the latter most native species can continue to survive in managed agricultural ecosystems (Huston, 1994). This implies the need for clarity of purpose in designing a project, in order to avoid an apparent dilemma between safeguarding biodiversity for use in the long term, or improving the well-being of local people in the short term (Adams & Thomas, 1993). Neglecting to explain to local people the special value of biodiversity in a reserve, and making it possible for them to respond to this knowledge, can result in its replacement by farmland even more quickly than if the project had never taken place (Oates, 1995).

10.4 Managing tourism

10.4.1 Categories of tourist

World tourism is a huge and dynamic industry, which barely existed before the 1960s, and which is subject to continually evolving fashions and interests. Carefully planned and managed tourism can have a positive role in conservation projects. Cultural, biological, and scenic features of the project area may attract visitors, who can be taught how to avoid doing harm, and used as a resource to support local aims. This can be hard to achieve safely, since the scale of global tourism is such that there are hundreds of millions of international tourist arrivals each year, and thousands of millions of dollars available for investment by the private sector (EC, 1990; Eber, 1992; Filion, Foley, & Jacquemot, 1994; Pleumarom, 1994). A new and interesting destination can thus quickly attract many visitors, and with them new pressures to develop facilities and services that can degrade local environments and cultures. This calls for care in planning and managing all aspects of tourism development.

General points include the need to minimize damage to local ecosystems and cultures, while ensuring that direct benefits accrue to local people. It is relatively easy for government officials and airline, car-rental, and hotel companies to obtain revenue from visitors. It can be harder for local people to benefit, even though they often have more influence in conserving or degrading the resource which attracts visitors. Conscious effort is therefore needed to ensure that local people receive adequate rewards for their co-operation.

One starting-point is to analyze the various kinds of visitors which might be attracted, and their different needs and expectations in relation to the resources of the project area. It may be possible to satisfy some kinds of tourist but not others, for given levels of investment and acceptable levels of impact, and this should be clearly understood from the outset. According to Cochrane (1992), the main categories of tourist are as follows:

- *Explorers*, who tend to be solitary and adventurous, and may be wealthy although they avoid non-essential expenses and require no special facilities.
- *Special interest tourists*, who are strongly motivated by a particular hobby, are willing to accept discomfort and long travel to satisfy it, and are willing to pay well to obtain necessary facilities and services.

- *General-interest tourists*, who may be keen on nature and safe but adventurous activities (such as white-water rafting and trekking), and who need good facilities, but will accept difficult conditions for short periods.
- *Back-packers*, who aim to travel cheaply for long periods, and may enjoy trekking through natural scenery.
- *Mass tourists*, who will accept high densities of other tourists, are interested in local culture and natural history if these are easy to experience, and generally want good facilities for accommodation and travel.

Explorers and special-interest tourists are the most likely groups to appreciate the special assets of a project area, to spend freely but constructively once there, and to be open to guidance on how to avoid doing damage. Explorers also have a special role in gathering information about a project area, which can be used to guide planning and investment. Special-interest tourists are the ideal group to use in most project areas, since densities are likely to remain manageable and expenditure per visitor can be quite high as long as the quality of the experience is maintained. Where the special interest concerns habitats or wildlife, the link with conservation can be direct and powerful. General-interest and mass tourists, by contrast, are less likely to help protect a nature reserve, and are more likely to damage it directly or indirectly.

Back-packers may also have a special role, if a project seeks initially to attract small numbers of visitors from among current travellers and local residents. They are more likely than others to have flexible itineraries, and to be influenced by local, word-of-mouth advertising. At Kanyang, in the Okwangwo Division of Cross River National Park in Nigeria, rudimentary but well-managed reception facilities attracted increasing numbers of local residents and back-packers, even without significant advertising (Caldecott, Oates, Gadsby, & Edet, 1990). In this case it was found that an important step was to agree fixed rates and conditions of service for locally hired guides and porters, including provision for hours worked per day, loads to be carried, food costs, and overnight charges. Many misunderstandings and arguments can be avoided by making an advance agreement with local people, and a written code of conduct for visitors.

10.4.2 Nature-oriented tourism

Many special-interest tourists are strongly motivated to visit natural areas where they can admire scenery and wildlife, and such people are known as

nature-oriented tourists. A subset of these are **ecotourists**, defined as people who use natural ecosystems without damaging them, while contributing directly to their continued protection and providing clear benefits to local people (Valentine, 1992). Few data exist on ecotourism which can be compared with overall tourism patterns, so they are usually pooled with figures on people who 'travel to enjoy and appreciate nature' (Filion, Foley, & Jacquemot, 1994). This description would strictly refer to the broader category of nature-oriented tourists, but authors differ in their use of these various terms (cf Boo, 1990; Hay, 1992; Pleumarom, 1994). The idea of ecotourism as defined above may be more useful in the future, since it highlights the differing roles of ecotourists and other tourists in contributing to the long-term security of nature reserves.

To encourage nature-oriented tourism beyond the level of casual visitors and back-packers, a project may need to invest in trail systems inside a reserve, including camp-sites and simple rest-houses, as well as in centres where visitors can be oriented and directed to sites, facilities, and services. Brochures, maps, and other materials can then be developed for the use and guidance of visitors, and for circulation to specialist travel agents to attract special-interest visitors. Cultural features can be integrated with a reserve's trail and visitor-information systems, as at Niah National Park in Sarawak, where an Iban longhouse on the edge of the park was treated as a stop in the trail system and explained in educational materials in considerable detail. This was consistent with the way this park was centered on the Niah Caves, which are an important archaeological site and a source of both edible birds' nests and guano used in growing pepper.

Rainforest nature reserves became steadily more important as travel destinations in the 1980s and 1990s, partly because of growth in global public awareness of the complexity of rainforest ecosystems. As visitors become more interested in plants and invertebrates, it becomes less important that it is difficult for casual visitors to sight large and spectacular animals in rain forests. Malaysia, Indonesia, Costa Rica and other countries have increasingly emphasized rainforest nature trails as additional facilities for tourists to visit, and economic analyses show that the use of rain forests for ecotourism can create greater value than any other use of the same land is likely to do (e.g. Tobias & Mendelsohn, 1991).

Progress in the tourism component of a project can be assessed by measuring the number of visitor-days spent in an area, the number of guide-days worked and the rate of pay, as well as by monitoring local public opinion about tourists and the satisfaction recorded by the visitors themselves. Because tourism is especially sensitive to the broader environment in which it occurs, investments should emphasize the management of whole systems. Thus, for example, a hotel on the Obudu Plateau, in the proposed Cross River National Park in Nigeria, was calculated to be an attractive stand-alone investment (Erokoro, White, & Caldecott, 1990). Sensitivity analysis of price and occupancy rates, however, showed that such an investment would be vulnerable to any deterioration in the area's attractiveness to visitors. According to visitor interviews, much of the Obudu Plateau's appeal depended on the quality of the whole landscape, so privatizing parts of the system could well have been counter-productive.

10.4.3 Diving tourism

Tourism in tropical marine locations grew rapidly in the 1980s and 1990s, especially in areas with warm, clear water where coral reefs tend to develop. The latter are comparable to rain forests in terms of their biological complexity, and virtually all visitors to coral reefs are interested in natural history or else become so with experience. Although coral viewing using glass-bottomed boats is widespread, and caters to the general-interest and mass tourist market, direct access to reef environments is mainly by snorkelling and SCUBA diving.

Divers who use SCUBA are almost the archetypes of special-interest tourists. They are strongly motivated by a hobby, bonded by common experience and specialist magazines, happy to travel in order to dive, and are willing to buy or hire expensive equipment and to rent local services such as expert guides, which implies a strong economic role. In Malta, for example, about 20 000 diving tourists visited the islands each year in the early 1990s, supporting 30 diving schools and spending about US$30 million locally (FUAM, 1994). At an individual level, a sample of 133 recreational dives in Indonesia in 1992–4 cost an average of about US$1.00 per minute spent underwater, including local travel, accommodation and guide charges (personal observations). For most divers, this willingness to pay depends on the clarity of the water, the abundance and variety of aquatic life, and the condition of undersea environments. Few enjoy diving over a sea-bed, for example, which is 'little more than a grey, lifeless ossuary' as a result of dynamite fishing (Pye-Smith & Feyerabend, 1994: 155).

Divers are thus deeply hostile to pollution and to destructive and indiscriminate fishing techniques, which are major threats to tropical reef environments. They will withdraw their custom from any location which has been damaged beyond a certain, rather low, threshold. This makes all SCUBA divers potential ecotourists according to the strict definition used above. Although most basic dive courses teach respect for undersea life (e.g. PADI, 1990, 1994), neglect of this aspect of training can lead divers to damage ecosystems. A dive site which receives more than 3000–4000 dives per year is likely to deteriorate (Schoen & Djohani, 1992; Dixon, Scura, & van't Hof, 1993). The impact of crowding can be reduced, however, by limiting access to experienced divers with good buoyancy control. A lack of proper mooring facilities and training for boat operators can also result in anchors being dropped upon, or dragged across, reefs. There are well-known ways to avoid such damage, and most divers are willing to co-operate with measures to protect the environment. It is harder to find ways for divers to contribute actively to the protection of dive sites and to provide clear benefits to local people, and thereby to become real ecotourists.

An example of one kind of solution is provided by the village of Ameth, near Separua in Maluku, eastern Indonesia. This community had an effective system in place in late 1992, under which divers from a nearby resort registered and paid a fee before diving there (Muller, 1992; personal observations). As a result, the reefs were continually supervised by the local people and were almost pristine, unlike many elsewhere which had been damaged by explosives used in fishing. Similar arrangements have been made in parts of Southern Luzon, Philippines, resulting in rapid recovery of reefs there (P. Guy, personal communication, November 1994).

It is important to remember that revenue from dive tourism is only one of many benefits which can be obtained by local people from the sustainable use of coastal ecosystems. All depend equally on avoiding open-access use of those systems, which means that local people must have the motivation, the physical ability, and the authority to exclude outsiders. Projects in the Cyclops and Cenderawasih areas of Irian Jaya have shown that motivation is easy to generate, provided that patrol boats are available and that the authority to patrol has been granted by government (Chapter 7). Dive tourism is one way to encourage communities to exert control over local resources, and to encourage government to help them do so.

10.5 Managing coasts and seas

10.5.1 Introduction

Most of the world's people live in coastal zones, which are vulnerable to the impacts of settlement, exploitation, and pollution generated within them, as well as to impacts created offshore or in inland catchment areas. The mixing of nutrient flows from land and sea, which makes these zones so dynamic, thus also exposes them to unique dangers (Johannes & Hatcher, 1986; Ray, 1988, 1991; Clark, 1992). Coastal and marine areas provide special challenges, and only a few points will be made here to guide project designers away from common mistakes.

Estuaries, sea-grass beds, mangroves, and coral reefs are all examples of coastal environments which support biological communities extending far out to sea. as well as human trade and subsistence patterns extending far inland. All can therefore be considered key ecosystems, and projects must protect them if people are to be supported within the project area and beyond. This means that nature reserves have a special role in marine and coastal conservation, which is less to do with preserving biodiversity than with maintaining productivity for human use.

Because of the difficulty of enforcing any law adequately over large areas of coast and sea, the attitude of local people is of paramount importance to the success of coastal conservation projects. As Laffoley (1995:15) observed about the protection of marine reserves in Britain: 'Local policing is probably the only way forward ... as central policing ... although necessary, can never cover all areas of the coast in sufficient detail because of the cost.' If this is true in Britain, it is far more so in countries such as Indonesia and the Philippines. The demands of marine surveillance differ greatly from those on land, however, because of access and travel constraints. This means that greater attention must be paid to increasing the physical capacity of local people to participate in conservation efforts.

10.5.2 Coastal ecosystems

Estuaries are among the world's most productive ecosystems due the large amounts of fresh water, minerals, and nutrients they receive from inland areas. Tidal currents mix and distribute these materials, leading to high rates of primary and secondary productivity. Estuarine habitats such as sea-grass beds, mudflats, and mangroves thus support a wide range of species and large numbers of finfish and shellfish. They provide important feeding grounds, spawning areas, and nursery sites for the larvae, post-larvae, and juveniles of both marine and riverine species, thereby linking these ecosystems. Juveniles of the commercially important tropical shrimp *Penaeus merguiensis*, for example, concentrate themselves in near-shore waters close to sites of freshwater discharge, before moving offshore as adults (Racuyal, 1994).

Mangroves are communities of plants which can tolerate life in tidally flooded salty mud. Mangrove trees often have air-breathing roots to resist waterlogging, stilt roots, tannin-rich vegetation, and bark to resist herbivore attack, and fruits which are dispersed by floating in sea-water. Their roots bind the mud together, and create a tough, tangled mass of woody material which can absorb the energies of wind and waves, thus protecting coastlines from erosion. They trap fallen leaves and silt, and are used as key feeding and breeding grounds by crabs, shrimps, other crustaceans, many finfish, scallops, cockles, clams, mussels, as well as birds, mammals, and reptiles.

According to Pye-Smith & Feyerabend (1994: 144), 'a study by the Asian Development Bank suggested that a hectare of mangrove swamp produces an annual harvestable yield of 100 kilos of finfish, 20 kilos of shrimp, 15 kilos of crabmeat, 200 kilos of mollusc and 40 kilos of sea cucumber. The same hectare also supplies an indirect harvest of up to 400 kilos of finfish and 75 kilos of shrimp which mature elsewhere – for example, on and around coral reefs.' It is thus common for mangroves to support traditional life styles based on small-scale or artisanal fishing, using small boats, intricate traps, and nets in shallow water. This is a very labour-intensive and productive use of the fishery resource, but one that is often underestimated because of bias in fisheries statistics towards landings by larger, commercial craft.

Undervaluing mangrove-dependent fisheries has often led to mangroves being seen as waste ground, resulting in many being converted to other uses such as private aquaculture ponds. The resulting loss of public fishing benefits can devastate the economies of coastal communities. This is aggravated by the loss of access to other mangrove products, of which over 70 have been documented worldwide, ranging from palm-sugar and honey to tanning materials, charcoal, wood-alcohol, and water-resistant poles (e.g. Hamilton & Snedaker, 1984). These uses can be supported by small-scale, continuous harvesting without seriously damaging the mangrove ecosystem or its fishery role. This depends, however, on the rate and pattern of harvesting, which depends on traditions of exclusive harvesting rights by local people. Where these are abandoned, clear-felling of mangroves for rayon or paper-pulp production has had very serious consequences for local and other people (Bennett & Reynolds, 1993).

Coral reefs are built by colonies of small animals (polyps), containing photosynthetic symbionts (zooxanthellae) which allow them to trap atmospheric carbon dioxide and to exude calcium carbonate to form solid structures in which to live. These structures comprise the reef itself, the solidity of which helps to protect coastlines from erosion as well as having an important global role in storing carbon. Coral areas are highly complex ecological systems, extremely species-rich and very productive. Reefs shelter hundreds of resident fish species, and many others use coral areas for spawning, for shelter as juveniles, or at some other stage of their life cycle (Myers, 1991; Sale, 1991; Kuiter, 1992). Thus, reef-dependent finfish and shellfish catches are a major source of dietary protein and revenue for coastal communities in areas where coral reefs occur (Johannes, 1981).

10.5.3 Marine reserves and fishery management

Most marine organisms are highly mobile at some stage in their life cycle, often by having larval planktonic phases during which they disperse widely (Angel, 1992; Peterson, 1992; Norse, 1993). This is an important factor for the design of conservation projects in marine locations. It means that, if habitats are protected from further disturbance, they can be recolonized from sources far away. It also means that, if key ecosystems such as coral reefs and mangroves are protected, they can act as sources of recolonization for other areas, while also supporting harvesting over large areas of sea.

This is quite different to the situation on land, where the productivity of artificial ecosystems is largely independent of that of natural ones within reserves. Plants from a terrestrial reserve are likely to be treated as weeds if they establish themselves in farmland, for example, whereas in the sea each planktonic fish larva is a potential fishery asset to be harvested after growth. Spawning and nursery grounds, migration corridors, and stopover points can all be vital for fish populations far off shore, and many of these are located in coastal zones. Because of these factors, there is a strong consensus that near-shore marine reserves are central to the sustainable management of coastal and marine environments and the wildlife populations they contain (Roberts & Polunin, 1993; Bohnsack, 1994; Russ & Alcala, 1994).

It is impossible to manage a fishery successfully by regulating the catch alone. This is because habitat quality determines the availability of resources to support feeding and spawning. Provided key habitats are intact, catching some members of a prey population frees resources which other members can use to grow and reproduce faster than they would otherwise do. This is the principle which underlies the concept of a maximum sustainable yield (MSY), in which just enough prey are caught to allow new prey to breed and grow with maximum efficiency, and with no deterioration of the system as a whole. To calculate such a yield requires detailed knowledge of the ecology of prey species and the dynamics of their populations, so it is much easier to do for simple, single-species fisheries than for complex, multi-species ones. In the latter, which are normal in tropical seas, data are usually scarce but each species is likely to have a different MSY, and each MSY will be valid only for a particular set of habitat conditions (Garcia, 1994).

Fish populations can be subjected to five main kinds of over-fishing (Lapidaire *et al.*, 1991):

- growth overfishing, in which fish are caught before they can grow;
- recruitment overfishing, in which the proportion of spawners in the population is reduced;
- economic overfishing, in which the effort per unit catch greatly exceeds that at a maximum sustainable yield;
- ecological overfishing, in which there is a shift from large, slow-growing species to small, fast-growing ones; and
- Malthusian overfishing, in which fishing effort is driven by competition among fishers.

In Maqueda Bay, off western Samar in the Philippines, the effects of all five kinds of overfishing were evident in late 1994, along with damage to benthic habitats caused by trawling, exploiting mud-living bivalves, and the use of explosives and poisons. Some of the original mangrove in the Bay had been converted to aquaculture ponds, and much of the rest was subject to firewood collection and charcoal making. These factors had undermined fishery productivity, so the various user groups were competing for access to a diminishing supply (Caldecott, 1994b). Overfishing resulted partly from open-access exploitation of the resource by large numbers of artisanal fishers, partly from the activities of commercial fishing craft (often using gear which was too effective, such as purse seines), and partly from a commercial trade in fish and fish products. The latter supplied the enormous market of Metro-Manila, and because of the capital-intensive nature of this trade, much of its value was being captured by middlemen rather than by fishers. This had deterred small-scale fishers from investing in more sustainable fishing techniques.

The commercial trawling business in Maqueda Bay developed in the years after 1980, when government incentives encouraged investment in the fisheries sector. In 1980–2, mean annual shrimp landings in the Samar Sea were over 1600 tonnes, of which about 5% were by trawlers. In 1983–7, however, mean annual landings had declined to about 1000 tonnes, but some 43% of the total was taken by trawlers (data in Racuyal, 1994). As well as showing the effects of over-fishing on the total catch, this indicates a shift in the flow of benefits in favour of trawlers, and at the expense of artisanal fishers (who use gill nets, fish corrals, push nets, and beach seines rather than trawls). In 1993, 107 trawling boats were based at coastal communities around Maqueda Bay (Racuyal, 1994).

With the implementation of the Philippines' Local Government Code since 1992–3 (Chapter 5), several municipalities around Maqueda Bay applied restrictions on trawlers. The Municipality of Paranas, for example, excluded trawlers within 7 km of shore, built physical obstructions under water, and enforced the law by impounding an offending boat and burning its nets. By late 1994, there was local public awareness of the link between intact mangrove and coral ecosystems, the productivity and diversity of fishery resources, and the prevention of coastal erosion (especially during typhoons, to which Samar is vulnerable). Several municipalities had established sanctuaries in coral and mangrove areas, and local people had noted the rapid return of fish species that had previously been lost (Tan, 1993).

Public expectation was that the exclusion of commercial fishing craft and recovery of key ecosystems would increase local food security and income levels. An active local environmental NGO grouping (the Maqueda Bay Ecosystem Forum) had helped create public awareness of these issues, and sought to draw attention to improvements as they occurred (MBEF, 1994). The overall impression in late 1994 was that the management of Maqueda Bay was at a turning point, with growing public support for excluding outsiders from municipal fishing grounds, banning destructive fishing methods, and establishing sanctuaries for key ecosystems. A very similar set of problems in the 1980s, and a near-identical process of solving them in the 1990s, were described by Pye-Smith & Feyerabend (1994) at the Municipality of San Diego and other communities around Talin Bay in southern Luzon.

These cases show how open-access exploitation can be a special danger in marine situations, and it has indeed been the main threat to worldwide oceanic fisheries for decades (Williams, 1994). Attempts to manage these fisheries have been thwarted by competitive exploitation involving fishing fleets from many countries, using modern techniques which are indiscriminate and extremely efficient, and whose effects are aggravated by pollution, bottom trawling and 'ghost' fishing with discarded nets (CEC, 1993). Fish and fish products are traded into a world market which is effectively infinite, and will accept any species in some form. The industry is heavily subsidized, poorly regulated, and weakly policed, there are few incentives to restrain catches, and there is little political impetus to solve these problems. Serious overfishing is therefore common and widespread, and has occurred in many areas with disastrous consequences for fishery-dependent local people (McKie, 1994; Lean, 1995; Ghazi, Smith, & Trevena, 1995).

The Law of the Sea Convention of 1982 came into force in November 1994, and provides a framework for negotiating the regulation of fisheries on the high seas (de Klemm & Shine, 1993). By 1995, however, a scientific consensus was emerging that global fisheries were in a more desperate state than had been thought (e.g. Beddington, 1995; Pauly & Christensen, 1995). This had begun to encourage some governments to take action outside the Convention to close down open-access oceanic fisheries. Taking the lead in early 1995 was Canada, which had been sensitized to the issue by the collapse of fisheries and coastal economies in Newfoundland and Nova Scotia (Ghazi, Smith, & Trevena, 1995). In this case, Canada was not seeking to establish exclusive rights, but rather to impose a complete fishing moratorium in parts of the northern Atlantic to allow fish stocks to recover if they could.

There are a number of ways to relieve pressure on fisheries, such as by agreeing quotas for fleets or individual craft, by banning certain techniques and classes of equipment, and by reducing the number of fishing craft through restrictive licensing, removing subsidies, buying out owners, or paying fishers not to fish. At a project level, the most important kind of solution to overfishing is often to limit access to fishing areas and to curtail fishing activity there. This implies the need to help coastal communities to use near-shore fisheries exclusively, and to strengthen the legal and practical basis for 'territorial use rights in fisheries' (TURF; Christy, 1982). Technical guidance may have a role, but establishing exclusive usage rights to particular water systems will usually be the key reform needed to safeguard wild fish production (Ruddle, 1989; Zerner, 1990; Ruddle, Hviding, & Johannes, 1992; Cooke, 1994).

There are two main ways in which TURF may be maintained, the first of which is also relevant to any form of fishery regulation. This is based on active policing, which requires visual surveillance, communications, fast boats, and the authority to order away or impound fishing craft, as well as the means to do so. The second is based on passive measures, such as underwater obstructors, which can entangle trawls and other nets used by large craft. Artificial reefs are sometimes built for this purpose, as well as to provide new habitats for fish, but these and other fish aggregation devices can have the counter-productive effect of making fish even easier to catch than before (Bohnsack & Sutherland, 1985; Brock, 1985; R. S. V. Pullin, personal communication, November 1994).

In conclusion, critical measures for conservation projects to take in coastal environments include the following. Firstly, there should be an emphasis on establishing and strengthening exclusive rights to harvest marine resources by local people, with whom a management régime can be agreed once TURF has been established. This will involve helping local people to obtain both the legal authority and the practical capacity to control access to fishing areas. Secondly, key ecosystems which are needed to sustain fisheries should be identified and protected as part of a locally agreed management plan. These will normally include mangrove and coral areas, but may also involve other coastal, estuarine, riverine, and inland watershed areas. Unless exceptional circumstances apply, a project should normally give top priority to maintaining TURF, and to preventing damage to these sanctuaries.

11

Options for changing people's minds

11.1 Introduction

Although wild species and people all have needs, only people are voters, investors, stakeholders, planners, and decision makers, and they may not feel inclined to care about wild species or to spend time and effort in conserving them. If so, then little is gained by studying the needs of wild species and 'speaking for the birds', or by advocating change in people's lives. Thus, conservation cannot only be about choosing where to put nature reserves, or how big they should be, or how many fish should be taken from rivers or deer from forests. Instead, it must involve changing people's minds.

The aim of doing so is to create a durable consensus in favour of conservation, expressed through choices in voting, spending, and consuming far into the future. Such a consensus would mean an overwhelming majority of people actively avoiding damaging nature and preventing others from doing so. This cannot be achieved against people's will, and nor can it be achieved quickly in any place where people are not already willing and able to conserve. It can be achieved eventually, however, and conservation projects should be designed with this aim in mind. They should therefore use procedures which help people make decisions about which they feel positive, and which are also ecologically sound. Some of these procedures are described below.

11.2 Tenure and incentives

Observations in Chapters 2–7 showed that the projects which had most local support, and which were most able to achieve conservation aims, were those which improved local people's security of tenure. This term is used here to mean people securely holding the means to satisfy their needs now and in the future, and controlling the resources which they need to determine their own future. Such improvements were effective whether applied to land own-ership, as in the Makiling, Cyclops, and Arfak projects; to fishing, hunting, and logging rights, as in the Cenderawasih, Wasur, and Ekuri projects; or to the control of financial savings and local business ventures, as in the Amisconde, Arfak, and Wasur projects. Strengthening tenure contributed greatly to the effectiveness of all these projects. By contrast, the biggest problems were seen in projects where people felt that their tenure was threat-ened. One example was the Cross River National Park project after 1992, where delayed and then partial implementation caused local people to fear the future rather than to have confidence in it, and thus to resist the efforts of conservationists.

An important link exists between security of tenure and confidence in the future, and both are likely to affect people's willingness to invest in any new undertaking, including conservation. This suggests that confidence is the key to changing people's minds about conservation, and that safeguarding tenure is a good way to achieve a durable mood of optimism among the people of a project area. A conservation project should therefore aim to help people feel more confident in a future containing high levels of biodiversity than in any other.

If people securely hold the means to meet their needs in the future, they may refrain from destroying biodiversity in the present. Whether they will or not depends on what those needs are, and how they are met. Human needs include those which are social, bodily, intellectual, spiritual, and emotional (Fig. 11.1). All are important, and all must be met if a person is to experience well-being. These different needs all interact, and the desired or achieved balance among them varies among individuals and over time. There is no way to quantify or compare them directly, but it is understood that all contribute to the concept of human development, both of individuals and of societies. The most relevant factor here is that opportunities to meet needs can be used as incentives which will encourage people to support conservation priorities. These incentives are as diverse and interchangeable as human needs themselves, but all are valid and should be considered as options for use by project designers (Box 11.1).

Box 11.1 Kinds of incentive which can be used to strengthen project effectiveness (adapted from Caldecott, 1992b)

Respect and self-respect. All incentives draw their effectiveness from improving respect and self-respect, but many societies value behaviour which involves damaging natural resources. A project may therefore need to encourage changes in how people win respect socially and feel they can earn self-respect.

Intellectual and aesthetic values. People who understand and are interested in nature and natural phenomena, or find them beautiful, tend to be against destroying them. Education can help to strengthen intellectual incentives, and to encourage new aesthetic standards where local perceptions of what is beautiful act against conservation.

Religious and cultural values. Where a religion encourages respect for nature, this can be strengthened through education (e.g. Abou Bakr *et al.*, 1983; John Paul II, 1989, 1990). Caution is needed, however, where conflict arises between cultural and conservation priorities, since cultural identity is often neither flexible nor easily divisible. Rapid change might therefore break down the very social institutions that are of greatest potential value to a conservation project.

Security of tenure. The knowledge that rights to resources are permanent and exclusive is often a pre-condition for managing them sustainably.

Money. The opportunity to obtain extra money can be a simple and powerful incentive. Tax based incentives, however, can only change attitudes amongst those who both pay taxes and understand the tax system.

Employment. Having a job means more than having a source of income, since it includes having a place and purpose in society, with attendant self-respect, duties to fulfil and interesting activities to perform.

Participation. People often enjoy doing things together, and they may therefore welcome the opportunity to participate in community discussions and activities.

Success. A small but unambiguous success in achieving some community aim can greatly increase confidence and motivation. For this reason, projects should start small with achievable pilot initiatives.

Laws. A law acts as an incentive only to the extent that it is understood and complied with in competition with other incentives, which may strongly encourage people to disobey it. Spending resources on removing competing incentives can be more efficient than spending them on enforcing the law itself. Voluntary compliance may be improved by education, provided the law is perceived as just, and by maintaining the habit of compliance among target groups of people through consistent enforcement.

11.3 Environmental education

Education is one of the most influential factors in economic development. It strongly supports economic growth and equity in income distribution, especially when investments are focussed on primary education, on economically weaker sections of society, in rural areas, and in developing countries (Tilak, 1989; Haddad *et al.*, 1990). Rates of return in educating girls and women are at least as high as those among boys and men, while also providing other social benefits such as reduced fertility and improved sanitation, family nutrition, and child health (Haddad *et al.*, 1990; Herz *et al.*, 1991).

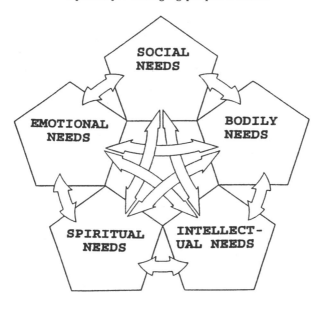

Fig. 11.1 Human needs

Education can thus be a potent tool with which to improve well-being, and should be a major feature of conservation projects.

Like any powerful tool, however, education should be used with care and awareness of side-effects, not all of which are benign. Casualties can include traditional belief systems, languages, ethnobiological knowledge, myths and legends, art forms, and local skills and technologies. These are vital resources for conservation, since they help people maintain confidence in themselves and in their societies, while also often supporting durable harvesting and farming systems. School-educated children may no longer wish to learn from their elders, and may lose confidence in their own society and its ability to provide a satisfactory life style. Young people may therefore drift away, often to cities elsewhere. Even literacy training can marginalize people if they are given only biased and limited vocabularies, making them less able to comprehend social, economic, and ecological realities, and to protect their own interests (R. J. Hammond, personal communication, August 1994). Lastly, all kinds of education and training can become instruments of oppression, if they are used to make people adopt or tolerate life styles which would not otherwise be acceptable to them (Freire, 1970).

On the other hand, literacy can improve people's access to new ideas, while public knowledge of nutrition, sanitation, and primary health care can reduce the amount of time and energy spent in sickness, in tending the sick, and in mourning the dead, thereby freeing people for other activities. Basic training in ecology and how the environment works will lay the groundwork for understanding how to improve farming and harvesting methods, and will make the ideas and ideals of conservation more accessible. These are all likely to improve people's ability to join in discussing and developing plans for their own participation in a project.

Most importantly, education promotes security of tenure, since without it people may be unable to detect and resist plans to take tenure away from them. A proposal to develop a fish-pond in a mangrove, for example, may benefit the family which owns it, but can turn many other people into tenureless refugees. If the potential victims did not know something of mangrove ecology, they would be unable to decide whether to resist the proposal, and would risk losing everything. Moreover, without education, people may not have the skills to use sustainably what they do hold securely. Secure tenure of an eroding farm with declining fertility is worthless, since it cannot meet human needs in the future. The same farm under a durable and productive agroforestry system, however, is an important asset. The investment needed to change one into the other will only happen with secure tenure, and if the knowledge to guide that investment is available.

Education depends on presenting relevant information in ways which the audience can understand easily, and people learn particularly well from others with similar backgrounds and problems. Success in environmental education therefore often comes more from networking and discussion among equals, than from formal teaching by 'high-status' teachers to 'low-status' rural people. Networking has the added value that it can promote solidarity, which also helps to promote tenure by allowing people to summon help or to organize with others to resist threats to it. It is important to remember, however, that the most effective networks are those which grow out of common interests and joint projects, rather than those which are designed and introduced by outsiders.

Education is an active process of guidance, but it should proceed in a certain direction (R. J. Hammond, personal communication, March 1995). It should start from an understanding of the **context** provided by the value systems of the people to be educated, then move to engage the **process** of improving thinking skills and confidence, and finally reach the **content** of what is taught. Trying to do it the other way round by emphasizing content first, as in a training course, risks failing to strengthen thinking skills and confidence. It will therefore leave people vulnerable to the influence of others, among whom may be people who wish to exploit their environment or their lives for profit. These may control television stations, newspapers, and other powerful media, and some of their messages may undermine people's tenure security. Education should help people detect such threats and understand their significance, and so help them defend themselves.

An aim of environmental education is to help people see how their own environment works, its importance to them and the consequences of their own actions while living in it. This should provide the skills which people need if they are to take part in decisions about how to use the environment. The environment should be thought of and taught about broadly, to include its biological, physical, social, historical, economic, and political aspects, as well as the government institutions and donor agencies (and conservation projects) which may influence it. This approach will help to establish a concept of the *total* environment, which contains many resources for use by people, as well as many constraints on what can be achieved within it. Environmental education will therefore strengthen linkages among all components of a project and its interaction with local people.

An important role of environmental education is to promote specific understanding of why biodiversity is important. This is needed because agroforestry and other farming techniques are available which can be used to replace natural ecosystems, without necessarily causing problems such as soil erosion even on steep and infertile land (Chapters 5 and 10). There is thus a need to emphasize the special values of natural ecosystems relative to all artificial ones. It can be stressed that biodiversity will buffer local farming systems against uncertainty, such as that caused by crop disease, bad weather, and climate change, and this can make the whole farming system more robust and secure. Such arguments can be joined with others which aim to help local people to see nature reserves as providers of essential services and useful options, rather than as unexplained green backgrounds to their lives (Box 11.2). It should be emphasized that many of these benefits can only be obtained fully if local people actively organize themselves to do so. They are opportunities, not hand-outs or free gifts.

Box 11.2 Reasons why people should welcome local nature reserves (adapted and expanded from MacKinnon *et al.*, 1986)

Advantage 1. Securing water supplies. Forests have root systems which cause rain-water to be absorbed deeply by the soil, and released gradually into streams, springs, and wells.

Advantage 2. Preventing floods and landslips. Forests smooth out the flow of water downhill, and their roots hold slopes together. Wetlands near rivers absorb excess water and protect farmlands and settlements beyond.

Advantage 3. Protecting soils and coastlines. Vegetation cover shields soils from wind and rain, and root systems bind the soil together to resist erosion. Mangroves and coral reefs absorb the energy of waves and currents, preventing the coast from being washed away.

Advantage 4. Moderating weather. Forests recycle water vapour and help keep the air moist and cool. They promote showers in the dry season, and moderate winds. Forests and coral reefs also store carbon, which would otherwise accumulate in the air as carbon dioxide and increase global climate change.

Advantage 5. Providing harvests within the reserve. Many wild animal and vegetable products can be harvested from natural ecosystems and used or sold. They include meat, fish, rattans, firewood, fruits, edible leaves, and medicinal plants. With care, they can be taken indefinitely from those parts of a reserve allocated to such harvesting.

Advantage 6. Providing harvests outside the reserve. Natural ecosystems are used for feeding, breeding, sheltering, migrating, and roosting by many species which can be harvested outside a reserve (e.g. fish, shrimp, and ungulates), and by others which pollinate crops or disperse useful seeds (e.g. bees, birds, and bats).

Advantage 7. Protecting genetic resources. Some wild species can be used to improve or diversify local farming systems, or else are related to crop plants and can be crossed with them to improve productivity or to increase resistance to pests and diseases. Many others contain chemicals which might be turned into profitable products, and local people may be able to benefit from these discoveries.

Advantage 8. Promoting tourism. Many outsiders may wish to visit a nature reserve, and local people may be able to use visitors as a source of money, jobs, and new ideas.

Advantage 9. Creating employment. Research, management, tourism, and other activities in and around a reserve will create many different kinds of jobs, from which local people can benefit increasingly as their training and experience increase.

Advantage 10. Opportunities for research. There is much to learn from studying natural ecosystems, and the more that is understood the more interesting and useful they will become to local people.

Advantage 11. Opportunities for education. A reserve provides many resources which teachers can use to improve education at local schools. Universities and high schools may want to use the area for teaching and field studies, and local people may be able to enrol as students.

Advantage 12. Opportunities for training. Research and management programmes involve people working in the area with new skills and technology, and many opportunities to learn from them will arise.

Advantage 13. Providing a clean and beautiful environment. Local people benefit from access to natural scenery and a lack of crowding and pollution, all of which can help provide a relaxing and enjoyable lifestyle.

Advantage 14. Attracting privileges. People who live near a reserve are stakeholders in its future, and may be eligible for special forms of assistance by government, NGOs, and others with an interest in conserving the area.

Advantage 15. Preserving traditional values. Traditions, cultures, and languages are all linked to the natural ecosystems near where people live. A nature reserve can help to preserve those links, and keep alive valuable local ideas and ways of doing things.

Advantage 16. Reasons for self respect. Nature reserves are special places, and as natural ecosystems become rarer they become even more so. People living near reserves are stewards of unique resources, and have a special and important role which is increasingly recognized by everyone else.

Advantage 17. Promoting self-determination. People who live within a viable ecosystem close to a nature reserve have a much broader range of options for development than most others. They can use their resources wisely to determine their own future without risking the loss of their means of survival and prosperity.

Advantage 18. Improving tenure security. A nature reserve can be used by local people to obtain legal rights to occupy or use nearby areas. This will mean agreeing to respect permanent reserve boundaries, but secure tenure outside will give more benefits than are lost by doing so. These can include clearer rights of inheritance, and better local ability to resist outside settlement or theft of resources.

11.4 Communication strategies

11.4.1 Public relations

Encroachment and over-harvesting are threats to nature reserves which will persist at some level for ever, since there will always be at least some people who are willing to exploit reserved resources for private benefit, or out of urgent private need. There are three linked means by which a reserve may be protected, these being vigilance, consistent enforcement, and public relations. Vigilance is needed because damage to a natural ecosystem can happen quickly, and recovery may take a long time. Consistency in enforcement is needed to maintain the reserve's boundaries in the minds of local people, since encroachment will happen only rarely if people are in the habit of thinking that the reserve is not available for other uses. This habit of thought is achieved only by consistent enforcement and reinforcement over time.

Public-relations skills are important in helping people to feel positive about the whole process of enforcement. Techniques include polite speech and respectful behaviour by enforcement staff, and firmness and fairness when enforcing the rules. These techniques and skills can be taught, and should

be taught to all personnel responsible for maintaining boundaries within project areas. They will not solve deeper problems, however, and land hunger
by poor people living next to a reserve will not go away just because protection staff are polite and consistent. Instead, public-relations skills have a role
in keeping people calm and receptive to ideas on how to solve underlying
problems. A relaxed and receptive atmosphere greatly assists, and can be
essential for consultation, planning, and negotiation.

In Sarawak in 1986, for example, a Penan community was badly upset
when one of their pet hunting dogs was killed by a truck belonging to a
logging company. This was an important issue to them, because of their
strength of feeling for the dog and also because the conflict represented to
the Penan the uncaring attitude of the logging company. Days of negotiation
over compensation followed and were finally concluded. As the logging company representatives drove away from the Penan group in their Toyota Land
Cruiser, however, they ran over another dog. This public-relations disaster
further focussed local hostility to logging, and contributed to the militancy
and discord over logging in Penan homeland areas which was described in
Chapter 2. If the outsiders concerned had represented a national park authority rather than a logging company, the public-relations issue would have
been the same, even if the deeper context was not.

Managers of conservation projects should be aware of other issues that
arose in several of the projects described in Chapters 2–7. The first of these
is that if benefits are to be given to villages early in the project, it is better
for small items to be delivered even-handedly than for large ones to go only
to a few communities. Secondly, each contented local employee is part of a
network of family and friends who will support the project, but this network
can turn hostile if an employee is fired or feels abused, so great care is needed
in devising a project employment policy. Thirdly, guidelines will be needed
to help project personnel react to issues like keeping wild animals as pets,
eating hunted meat when staying at villages, and allowing hunting for the
pot during field work. Fourthly, guidelines may be needed for presenting the
project to local people with caution and without exaggeration, since describing its benefits otherwise may lead later to the feeling that promises have
been broken. Lastly, special care is needed where the issue of resettling a
community may arise. This is a sensitive subject which must be approached
only through full consultation and careful planning. It is not one to mention
casually during a village meeting.

11.4.2 Information services

Providing accurate information about the project to local people is an important way to start building awareness and confidence. One way to do this is through dialogue in informal settings, in which curious people talk with others whom they know and trust. An example is provided by the role of the village liaison assistants in the Cross River National Park (CRNP) project in Nigeria (Chapter 3). This is especially important where illiteracy limits people's access to information in written form, but can also be used to supplement other approaches. The latter include the use of project newsletters, an example of which is given in Fig. 11.2, again from the CRNP project (Siakor & Ashton-Jones, 1992). Local thinking skills will be enhanced if people produce their own educational materials rather than importing them from elsewhere. Locally produced newsletters and radio shows can be cheaper and much more effective ways of promoting environmental education, than TV broadcasts by people far away with different priorities. They are also more likely to be focussed on issues which are relevant to people's lives. It is therefore best for local people to prepare their own educational materials, at times using information provided by the project.

If posters, leaflets, and newsletters are produced by others, they should be carefully researched and pre-tested to make sure they can be understood by the target audience and that they convey the intended message. A fund-raising brochure produced by an international NGO in 1988, for example, was aimed at potential donors who held financial assets in African and Latin American currencies as 'blocked funds', which meant that they could not be converted into other currencies. The idea was that holders of blocked funds might donate them to the NGO's projects which needed local currency in the countries concerned. Distribution of the brochure was cancelled, however, when a final check noticed that it was sub-titled 'How to put your worthless money to good use'.

The impact of information materials should also be assessed regularly, and care should be taken to avoid raising expectations too quickly. Opportunities may arise to increase the efficiency with which leaflets are distributed, and a project should be ready to respond to these. Living Earth in Cameroon used beer trucks to deliver newsletters to rural teachers, for example, mainly because beer was the one commodity which reliably passed through police roadblocks during times of civil disturbance (R. J. Hammond, personal communication, 1994). In Sarawak, leaflets explaining the law on protected animals were sent to District Offices in time for the annual renewal of shotgun licenses in 1987. This took advantage of the fact that most gun owners would

Fig. 11.2 Say No! to endangered species' meat

make their way to town at that time, if at no other, and had strong reasons to renew their licenses, which cost little and allowed ammunition to be bought legally.

11.5 Influencing decision makers

The ability to influence decision makers is essential to conservation projects, and there are several factors which should be considered in trying to do so. It is hard to influence hostile people, so there is a need to avoid offending powerful individuals or interest groups. On the other hand, a project area may already contain a number of such groups, each with its own priorities which may conflict with some conservation aims but not with others. There is also the question of how to present the case for conservation in such a way as to win approval and support.

One starting-point is to avoid giving offence to anyone, at least while the social and political circumstances are being researched. This calls for flexibility in approach and presentation. A project area, for example, may be divided into administrative units which are controlled by existing interest groups. Rather than working only with these units and groups, the project might opt to define itself using ecological boundaries such as watersheds. This will take the project out of the current political framework, and may allow it to proceed without interference. This approach was used by Living Earth in an environmental education project in the Lago de Valencia region of Venezuela (R. J. Hammond, personal communication, February 1995). The project area was defined using watershed boundaries, which were seen as politically neutral and thus non-threatening to the various local administrative powers.

A project may also respond to opportunities created by new circumstances. One example was the division of Cross River State in Nigeria in 1987, which increased the influence of rural people and helped the State Government to support a conservation project (Chapter 3). Another was political change in Samar and elsewhere in the Philippines in the early 1990s. In this case, a period of intense polarization was then coming to an end, as traditional political distinctions between 'left' and 'right' became obsolete, and this allowed former opponents to work together within environmental NGOs (E. C. Piczon, personal communication, November 1994). Conservation work was increasingly based on community-based resource management, including territorial use-rights in fisheries, and tended to favour the interests of poor rural people over those of large traders and urban groups. The process was helped by the implementation of the Local Government Code, which empowered local communities but did not consistently favour any established political faction (Chapter 5).

More active ways for projects to obtain influence include being aware of opportunities to help other agencies meet their own objectives. A project may, for example, help a government agency to define its needs for technical advice or training, and then offer services which meet those needs. This can be applied at a practical level, where agency staff need to know how to manage ecosystems which are unfamiliar to them, or to apply new group facilitation techniques, or else to use new technologies such as computerized GIS, GPS, and database units. It can also be applied at a policy level, since governments often endorse international policy documents before all their staff fully understand them. Some agencies will therefore accept the need for training based on such documents as the *World Conservation Strategy* (IUCN, UNEP, & WWF, 1980), *Our Common Future* (WCED, 1987), *Caring for the Earth* (IUCN, UNEP, & WWF, 1991), *Global Biodiversity Strategy* (WRI, IUCN, & UNEP, 1992) and *Agenda 21* (UNCED, 1992). This may also apply to the international conventions on climate change and biological diversity (de Klemm & Shine, 1993; Glowka *et al.*, 1994), and to multilateral funding initiatives such as the Global Environment Facility (GEF, 1994). Giving prominence to sustainable development policy can help to strengthen the case for action in any country, since it shows that official endorsement for conservation is widespread.

Another approach to advocacy is to try to make sure that the project's aims are explained clearly in terms likely to be of particular interest to decision makers. The aim of much of the work described in Chapters 2–7, for example, was to help powerful people to see the environment and biodiversity as specific and valuable, rather than as vague and valueless, in the hope of influencing their decisions. Much similar work has been done by others, and as a result donor agencies and governments have become more sympathetic to the idea that damaging environments has economic costs which affect overall economic performance (Munasinghe & McNeely, 1994; Turner, Pearce, & Bateman, 1994). It is now widely realized that if national accounts fail to record damage to stocks of natural capital, they will allow hidden costs to accumulate which will undermine the national economy (Repetto *et al.*, 1989; Repetto, 1991). These capital stocks include forests, fisheries, watersheds, fresh water, clean air, and fertile topsoil, as well as the biodiversity associated with them. This awareness can be used to justify public investment in preserving wild species and natural ecosystems.

Advocacy can be focussed on biodiversity in particular by emphasizing the total economic value of its components. Contributing factors are direct and indirect use values, and option, bequest, and existence values (Barrett 1988; McNeely 1988; Ruitenbeek 1990b). The last three are highly subjective, since they reflect the fact that people want to keep the option to use wild species for themselves or their descendants, or that they want a natural feature to be preserved even without wanting to use it. People who give option, bequest, and existence values to biodiversity tend already to favour conservation, so these values are less important than use values in advocacy. Use values are much less subjective, and are both more helpful in advocacy work and more likely to feature in the economic cost-benefit analyses of project proposals (Chapter 8).

The direct use values of natural systems depend on their commercial, non-commercial, and subsistence use, whether or not this is for consumption. They are the easiest values to quantify, since harvests can be costed at market prices and access and use fees can be counted. Indirect uses of biodiversity occur where natural ecosystems support economic activity and protect property. Examples include floodplain ecosystems recharging aquifers (Chapter 8), mangroves and coral reefs sheltering coasts and supporting fisheries (Chapter 10), and forests controlling floods and erosion and regulating global climate by storing carbon (Box 11.3). The latter, whole-ecosystem arguments can be used to justify national and/or global public investment in protecting natural forests. In the case of carbon storage, however, estimates of forest and coral-reef values can be hard to interpret because of uncertainty over the models used to calculate them, and over the division of costs and benefits between the country concerned and the rest of the world.

Box 11.3 Whole-ecosystem economic arguments for conservation (from Magrath & Arens, 1987; IPCC, 1990; World Bank, 1991a; ADB, 1992a)

Soil erosion. The rate of soil erosion from beneath an undisturbed tropical rain forest is typically around $20 \, t \, km^{-2} \, year^{-1}$. This increases by over 1000% during and immediately after selective logging, and returns to the pre-logging rate gradually over the next ten years, resulting in a cumulative loss of soil of about $500 \, t \, km^{-2}$. These figures do not apply where forest is exploited on very steep slopes, or where re-entry logging prevents regeneration. Repeated logging or farming, and associated fires, can degrade rain-forested land to grassland and scrub, with erosion rates of about $4000 \, t \, km^{-2}$ over ten years.

Off-site economic impacts. Soil erosion and hydrological problems have numerous downstream impacts such as damage to dams, ports, coral reefs, and fish stocks by sedimentation, and damage to farmland and infrastructure by flooding. Many of these can be measured, although their economic impact will be specific to each watershed. One estimate for the cost of watershed degradation in Java, Indonesia, suggested an annual indirect use value for intact watershed forests of about US$3000 km^{-2} year^{-1}.

Carbon storage. Both tropical forests and coral reefs contribute to the global distribution of carbon among vegetation, water, and the atmosphere. Increasing atmospheric carbon dioxide is thought to be largely responsible for global warming through an enhanced greenhouse effect. Tropical rain forests contain 17 500–25 000 t of carbon km^{-2}, and destroying them releases this carbon through burning and decay of vegetation. In terms of the marginal cost of achieving global goals in reducing carbon emissions, the value of rainforest carbon storage has been estimated to be about US$28 t^{-1}, although lower figures have also been used. The global carbon balance opportunity cost of logging tropical rain forest has been calculated to be up to US$190 000 km^{-2}, and the cost of converting it up to US$380 000.

Indirect use values of biodiversity are less easy to quantify than direct ones, since biodiversity itself is often only a contributory factor in any environmental service, and each service may contribute only part of the value being measured or estimated. In making an economic case for conservation, the values of wild species should therefore be emphasized separately from the services performed by ecosystems. Thus wild species may be used commercially to help pay for conservation or to support economic development. This process might use whole organisms, or else the chemical richness of ecosystems such as tropical forests and coral reefs, which include many materials with industrial potential.

This is an argument which bridges use and option values, so it is useful where an audience has enough background knowledge to understand the potential rewards of conservation. Some medicines derived from tropical forests were listed in Chapter 10, and suggest their direct use value as sources of future products. In making this point, it is necessary to explain how potential products can be found, developed, and marketed in a way that will benefit the constituents of the decision maker concerned. It will therefore be helpful to refer to the experience of Costa Rica in managing its national biodiversity inventory and protecting its national interests (Chapters 6 & 8).

The possibility of obtaining revenues from tourism has often been used to motivate governments to create nature reserves and to invest in their management (Chapter 10). The rapid growth in world tourism, and in nature-oriented tourism during the 1980s and 1990s, means that most governments are already convinced by this argument. There are still cases, however, where conservationists may see that a reserve or ecosystem has tourism potential, or is being used by tourists, before this is perceived by government. Under these circumstances, government decisions will be made in a state of ignorance of the economic significance of the resource, and natural ecosystems may be damaged as a result.

An example is diving tourism in Indonesia, which grew quickly in the late 1980s and early 1990s and supported the growth of dive resorts and specialist travel agencies (Muller, 1992). Despite this, many reefs in Indonesian waters were badly damaged over the same period, including those in the Pulau Seribu and Bali Barat National Parks off Java and Bali respectively, and many others in eastern Indonesia (Chapter 7). The country had committed itself to create about 300 000 km^2 of marine reserves by the year 2000 (Chapter 4), but the scale of damage occurring in the 1990s showed that more investment in protection was needed, both by government and indirectly, to support action by local people. Neither the total number of potential dive sites in Indonesia, nor the contribution of dive tourism to overall tourism had yet been documented by the mid-1990s. The lack of such economic data made it hard to use diving tourism as part of the case for increased investment.

The usefulness of such information is shown in Malta, where data compiled in the mid-1990s established that diving was a far more important part of Malta's tourism industry than had previously been thought (Chapter 10). This prompted the Maltese Government to expedite proposals for a number of marine reserves (B. Ferrante, personal communication, December 1994). Conservationists should thus be alert to the opportunity to gather data on the economic role of the ecosystems they wish to protect. Useful data are often accessible, and can greatly improve government willingness to invest in conservation if they are presented well to decision makers.

11.6 Building agreements

11.6.1 Managing conflicts

Conservation problems involve conflicts among people with different views on using resources. Projects to relieve such problems should therefore contain forums where people can talk with each other and negotiate peaceful solutions. The techniques of principled negotiation (Fisher & Ury, 1981) are available to help bring people into forums and to assist in finding fair and durable agreements once they are there. Other techniques can also be used in support of this process once a viable forum has been established (e.g. Krumpe & McCoy, 1992; Lewis, 1992, 1993; Bleakley, 1994).

It is hard to resolve a conflict before it is understood, and to do this requires careful analysis of the disputed resource, and of the concerns, needs, and priorities of the people with something to gain or lose in the dispute. Conflicts can be classified as being due to misunderstandings, exclusive or non-exclusive differences of priority, or differing values based either on common concerns or on clashes of ideology (Box 11.4). Conflicts based on exclusive differences in priority often involve rational positions on substantive issues, and are resolved either through power struggles or through study and discussion followed by sharing or compensation. Solutions usually mean that all sides lose a little, but are then able to participate in a more harmonious overall system. Many such conflicts are made worse by differing values, especially where there is no forum in which the people involved can interact directly. Not all conflicts based on differing values are equally insoluble, however, or can be solved in the same way. Conservation projects often involve numerous examples of all kinds of conflict, some of which may evolve from one kind into another as precedents are set and the positions of interest groups become entrenched.

Box 11.4 Conflicts of interest (adapted from Bleakley, 1994)

Misunderstandings. These arise from ignorance, and harm may be done to environments and to other users by accident and lack of care, which can be corrected by education and training. Examples include the damage done to coral reefs by the anchors of boats carrying parties of divers, and the trampling of fragile plant communities by tourists.

Exclusive differences of priority. These arise over resources which can be used only in one way or another, but not in both. They are rational disputes over measurable benefits, and can be settled by a clear decision in favour of one group of beneficiaries or the other, or else by sharing. Examples include conflicts between farmers and nature reserves, between owners of large fishing craft and users of community fishing reserves, and between private investors in aquaculture and artisanal fishers around mangroves. Settlements might involve education and compensation, for example by providing titled land or an uncontested concession elsewhere.

Non-exclusive differences in priority. These arise when people want to do different things in the same area using different resources. These conflicts are typically avoidable or soluble through compromise and design effort. Examples include disputes between people who want to catch crabs in a mangrove and others who want to harvest firewood there, between tourists and fishers who want to use a coral reef, and between scientists who want to study wild species in a forest and people who want to collect rattans there.

Different values due to a common concern. These arise when groups share common priorities, but differ on how to achieve their aims. Improved education, information flow, mediation, and leadership can help. Examples include conflicts where one group wants to exclude trawlers using gear obstructors while another side wants to strengthen legal territorial use-rights in fisheries and use patrol boats instead, and where one group wants to use NGOs to pressure local government while another wants to get elected to local government and work from within the system.

Different values due to a clash of ideology. These arise when groups have different basic principles which may be mutually exclusive. They may be solved by 'agreeing to differ', but different values may continue to generate conflicts which will need conciliation or arbitration. Examples include conflicts where one group believes in converting natural capital into money and investing it elsewhere while another group believes in sustainable use of renewable natural resources, and where one group holds that wild species exist only to serve people while another group believes that all species have a right to exist for their own sake.

Conflict management involves preventing, reducing, resolving, or arbitrating conflicts, and its procedures either aim to avoid harmful impacts on other people or to mitigate or compensate for them. Conflict resolution is a different process, which aims to achieve a satisfactory outcome for all parties, often through conciliation, mediation, compromise and effort in designing alternative solutions (Bleakley, 1994). Conflict resolution is not about deciding who is right or who is most powerful, but is about finding a satisfactory solution. The process of conflict resolution is begun by bringing parties to a dispute into a forum, where they can communicate with one another. This requires a degree of trust to be established and maintained, and there are four basic rules to be followed at every stage:

- listen, and hear;
- show that you have heard;
- say what you will do; and
- do what you have said.

11.6.2 Group facilitation

An important part of the conflict management process involves increasing social cohesion and co-operation within the target community. A number of group facilitation techniques have been developed to promote this (Box 11.5). These techniques owe much to methods developed in management training and in participatory rural appraisal. The latter comprises an important class of techniques which people can use to document the life style of their community, to assess problems, and to evaluate solutions. Chambers (1994a) provides an introduction to the various techniques involved.

Box 11.5 Participative extension techniques (adapted from SMAP, 1993b)

Group facilitation techniques generate and direct discussion. Examples include: *scanning* (obtaining a quick response to an issue from all participants); *buzz sessions* (all participants briefly discuss an issue before giving a quick response to it); *question and answer* (questions are asked and responses recorded); *jigsaw* (participants assemble fragments of a picture showing some aspect of their environment); *living photo* (groups act out a picture and freeze); *mini theatre* (groups act out a story); and *role-play* (participants adopt roles and interact in character).

Icebreakers and focussing techniques develop workshop spirit, introduce participants and focus attention on a given topic. Examples include: *group shape* (groups co-operate in using their bodies to form a familiar object or shape); *name chain* and *zip-zap* (participants take turns to call out each other's names quickly); *according to* . . . (teams arrange themselves quickly by height, birthday, etc.); and *agree/disagree* (dominant issues and personalities emerge from the group's response when confronted by ambiguous but controversial statements).

Identifiers are techniques which help participants identify topics, issues, questions, problems, and solutions. Examples include: *force field* and *problem/resource analyses* (groups draw pictures showing forces, or list factors which may help or hinder their achieving certain goals); *vision setting* (groups make and discuss collages, paper shapes, or symbols to represent their vision of development); *card display* (participants write or draw single words or images on cards in response to a key question, which are then displayed and discussed); and *what if* . . . (groups report on how they would approach certain problems if, for example, they were completely cut off from the outside world).

Analyzers are techniques which help participants analyze topics, issues, questions, problems, and solutions. Examples include: *self awareness* (participants describe their own current lives and how they would like to live in five years, and then repeat on behalf of another participant); *problem-prioritization* (the group agrees criteria for choosing priorities, and then seeks consensus on the priority problem facing them); *card stacking* (participants choose among cards representing candidates as the most serious problem); *web analysis* (each group draws a web of causes and preconditions connected to a problem or goal, which is then explained to the meeting); and *tree analysis* (each group draws a tree whose roots represent problems and its branches their consequences).

Deciders are techniques which help participants make decisions. Examples include: *decision-making exercises* (e.g. teams draw and discuss a map of who they relate to, work with, or try to influence in real life; groups make a real decision and explain how they made it; or individuals and teams

separately answer questions on how they would survive in a wilderness, and
compare the results); *round-the-world trip* (each group has one ticket which
only one member can use and which cannot be cashed, and must reach a
consensus on what to do with it); *debate* (groups select speakers to argue for
or against a position); and *chaired discussion* (a participant is nominated or
elected to chair a meeting, and the community is placed in control of a
discussion).

Team building techniques promote team spirit. Examples include: *table lift* (each
team holds a table at arm's length with arms straight and horizontal until they
tire, when they change the way they hold the table but without letting it touch
the ground); *chair carry* (participants try to carry an occupied chair, first one
by one, and then with others); and *construction game* (groups work together
to create a model quickly of anything they choose, using materials provided,
with criteria for winning specified in advance – e.g. the 'tallest' or 'most
imaginative' object). Community dances, such as *reels and square-dances*,
can also be effective team-building activities for use after hours.

*Behavioural techniques help improve communication, thinking, and social skills
of various kinds.* Examples include: *active listening* (in each group of three,
by rotation one talks about something emotive, another then paraphrases the
speech, and the last then summarizes the feelings expressed by the first); *not
listening* (each member of a pair plays a role and expresses opinions which
are incompatible with the other's, while not listening); *dialogue or monologue*
(a play in three scenes is performed to illustrate the Côte d'Ivoire saying that
'he who talks incessantly, talks nonsense'); *listening in silence* (each member
of a group takes it in turn to talk on a controversial subject while the others
listen); *listening in pairs* (pairs discuss a controversial subject, but after each
has spoken the other may not comment before summarizing what has just
been said to the speaker's satisfaction); and *problem tangle* (volunteers representing government and NGO workers leave the room, while all other participants join hands and form a human tangle to symbolize a community problem; the volunteers return and help disentangle the community, and this is
repeated with a more difficult tangle but without the help of the volunteers).

11.7 Open planning

11.7.1 Public involvement

It is better to avoid conflict than to have to resolve or manage it. Avoidance
depends less on anticipating all of the problems that might occur in a project,
than it does on involving in the project all individuals and the groups and
interests they represent. The aim is to create a structure in which people can
discuss issues before they become problems. The more involvement is
allowed for, the less likely the project will be to generate conflicts, and the

easier it will be to resolve them (Bleakley *et al.*, 1994). Local people may be involved at one or other of five different levels:

Level 1: Where local people are completely excluded.

Level 2: Where there is distant interaction only (e.g. through information handouts).

Level 3: Where there is some dialogue (e.g. through meetings with officials and consultants).

Level 4: Where there is considerable participation or partnership (e.g. through local NGOs).

Level 5: Where local people are completely in control.

Each level of involvement carries with it both opportunities and dangers. Very low levels of involvement can allow projects and 'paper parks' to be created very quickly, but can lead to alienation, dependency, and hostility among local people. Intermediate levels of involvement can achieve more durable results, but may be unable to create enough sense of local 'ownership' for the project or nature reserve to survive once external subsidies are withdrawn. Full control by local people is most likely to be socially durable, but risks local priorities obscuring the broader, long-term, public-interest and global-interest reasons for conservation. Conservation projects should therefore be designed to achieve as much local participation as possible, but without losing the opportunity to protect the interests of non-local communities and future generations.

This means ensuring three things. Firstly, a project should possess a forum in which all stakeholders can be kept aware of each others' activities, plans, and needs. These stakeholders should include accountable representatives of local people, interest groups and NGOs, agents of the national conservation agency and other government institutions, and agents of donor groups and international NGOs. Secondly, this local forum should be able to communicate with other levels of government, both to introduce their ideas for discussion, and to request help in solving problems that cannot be solved locally. Lastly, the field activities of the project should be driven by open planning, a process in which people are encouraged and enabled to become involved at every stage of a project from its initial conception to its final evaluation.

11.7.2 An open planning model

Each project will have a slightly different approach to working with local people, according to environmental conditions and the changes which are needed in the project area. Many projects have the common theme of trying to improve the economic well-being of local communities while encouraging them to make better use of resources. Such changes will be most effective if they are made in response to needs perceived by local people, and if they are introduced in ways which promote self-reliance and discourage dependency. A general model for introducing such changes in a way that is compatible with conservation is outlined in Box 11.6. It is based on methods used by European Union-funded projects in the Philippines, especially the Southern Mindanao Agriculture Programme (SMAP, 1992a, 1992b, 1993a, 1993b) and the Western Samar Agricultural Development (WESAMAR) Programme.

The basic idea is to use an open but structured planning procedure to achieve high levels of participation and concrete results in the form of 'micro-projects'. The aim is to help communities become self-sufficient in analyzing problems and in designing, implementing, and evaluating micro-projects to solve particular problems. Each community is encouraged to review progress regularly, and to prepare a community development plan (CDP) which is updated annually. The CDP should explain the development priorities of the community itself, and provide a context for subsequent micro-project proposals. The latter will be screened by the project before being supported, using criteria according to its own aims. Progress is measured mainly through the preparation of CDPs and the completion of successful micro-projects which are consistent with the CDPs.

Box 11.6 A development process for introducing change through open planning (adapted from SMAP, 1993b)

Stage 1: Entry. Based on its own priorities, the project selects a community with which it hopes to work. It then makes contact and seeks permission to proceed. After initial discussion with local leaders and distribution of information about the project, a more inclusive meeting is held at which the project and the development process are explained in detail. The community is asked whether it wishes to participate, and if so an entry agreement is signed to that effect. This agreement confirms the intent to collaborate, and explicitly states that project input is not to be permanent.

Stage 2: Trust-building. This begins with a community-wide meeting which identifies, analyzes, and prioritizes the aspirations of its members and the obstacles that may prevent them being achieved. Further meetings are planned to focus on finding ways to overcome each problem. These problem-specific meetings analyze the causes and consequences of each obstacle, and investigate a range of solutions, from which the community chooses one and develops a plan of action to implement the solution. This plan of action forms the basis for the first micro-project, which is implemented jointly by the community and the project. On completion, the micro-project is evaluated by a community-wide meeting, which assesses what has been learned from the process.

Stage 3: Learning by doing. The aim is to help as many community members as possible to learn how to design, implement, and evaluate micro-projects, through direct involvement in several micro-project cycles. This stage begins with the preparation of a community development plan (CDP) to give a context for subsequent micro-projects. The community is encouraged to make such a review and planning exercise an annual event, so that the priorities of each year's CDP can be incorporated into the project's own workplan and budget for the coming year. Priority micro-projects should not be planned in detail in the CDP, since later problem-specific meetings will develop each micro-project proposal. Special-interest groups will emerge during this stage, each with its own ideas for micro-projects to meet their own needs. The project may operate with these interest groups and with the community as a whole on several micro-projects at the same time. During this stage, responsibility for all aspects of micro-project design and implementation is steadily passed to the community.

Stage 4: Consolidation and networking. The aim is to help the community to see itself as a unit, which may be made up of different interest groups but which can share goals and work together to achieve them. The stage begins once the community has largely taken responsibility for the micro-project process. The key activity is a review meeting at which the various interest groups are recognized, the structure of the community is documented, and discussion focusses on shared attitudes, leadership arrangements and relationships with the outside world. The CDP will be revised, and used as a basis for networking with other communities elsewhere, both within and outside the project area.

Stage 5: Demonstration of capability. The aim is for the community to demonstrate to itself that it can successfully manage its own development. It is therefore conducted entirely by the community, and the project has no role other than as an observer. The community undertakes a review and planning exercise and implements micro-projects using its own resources, or else by accessing external resources other than those of the project. The project first agrees with community leaders that local teams have sufficient skill in data-gathering, planning, training, supervision, monitoring, and evaluation, and that the project is no longer needed and should move on to help others elsewhere. Local leaders should facilitate a review and planning meeting to help everyone realize what has been learned from previous micro-project cycles, before announcing the impending departure of the project.

Stage 6: Exit. The aim is to help the community accept that direct involvement by the project is no longer needed, and to exchange feedback on the whole experience between the project and the community. Both sides should help the other see the strengths and weaknesses of their procedures and plans. Once evaluations have been exchanged and discussed, an exit agreement is signed to end the project's formal relationship with the community.

It is tempting for outsiders to see a community as a single and homogeneous entity, since this greatly simplifies some of the ways in which a project might interact with it. Such an approach, however, would ignore the fact that all communities are internally stratified and that their members usually have different access to resources according to gender, age, socio-economic status, and sometimes also ethnicity (Li, 1993). Recognizing this diversity is essential if a project is to be inclusive and fair, and if it is not to provoke conflict by accidentally strengthening certain groups at the expense of others. The model described in Box 11.6 goes some way towards recognizing diversity by allowing for interest groups to emerge within a target community, and for the project to develop activities with those groups as well as with the community as a whole. Other ways to increase the project's sensitivity to local diversity include the study methods summarized in Box 11.7.

Box 11.7 Guidelines for studying communities without losing sight of their internal diversity (adapted from Li, 1993)

Guideline 1: Collect data on activities, ownership, beliefs, and knowledge in a gender disaggregated form.

Guideline 2: Collect data on the gender division of labour, seasonality, access to resources, and control over resources.

Guideline 3: Use maps and diagrams to show access to and control over resources, sources of livelihood, and men's and women's seasonal activities on and off the farm.

Guideline 4: Focus on what men and women actually do, with emphasis on the variation in female and male roles.

Guideline 5: Use life history and other techniques to determine changes in gender relations over time.

Guideline 6: Use direct observations to back up and cross-check interview data, in order to avoid bias in reporting.

Guideline 7: Train field researchers not to take 'no' for an answer, so that they will explore further when told, for example, that a woman is 'just a housewife'.

Guideline 8: Give special attention to women who head households due to male migration, death, divorce, or abandonment.

Guideline 9: Be alert to the way words are used either by community members or researchers to describe male and female roles and activities.

Guideline 10: Keep socio-economic status in mind, and find ways to allow different groups to express their concerns directly, rather than through representatives.

Guideline 11: Recognize that women and poorer men may not speak out freely in community meetings, and that other techniques may be needed to find out their views, problems, and priorities.

Guideline 12: Avoid hasty identification of needs and priorities which may over-simplify or distort issues, and recognize that some awareness-raising may be needed before needs can be identified.

Guideline 13: Keep the community fully involved as partners in data collection and validation of results.

Guideline 14: Use with caution tools (such as agro-ecosystem analysis) which rely on secondary data and community meetings.

Guideline 15: Use social and spatial units for analysis and problem definition which are as small and homogeneous as possible.

Guideline 16: Use approaches which involve people defining their own data needs, collecting and managing data themselves, and using their data as a resource for planning their own actions.

Guideline 17: Trust key informants, observations, and qualitative interview data as sources on many issues, and use surveys only where quantification is strictly necessary.

The open planning model described in Box 11.6 will at least prevent mis-understandings such as those illustrated in Fig. 11.3, but it will not necessarily promote conservation unless environmental education is also used. Any environment is limited in its ability to deliver resources with which to improve the economic well-being of local people. These limits are especially narrow in tropical areas with harsh climates and fragile or infertile soils, which is where many conservation projects occur. Such projects should try to help people understand and accept the limits imposed by their environment, as well as its potential for sustaining improved well-being. To do this requires that people are encouraged to identify and describe which aspects of well-being are most important to them, while also expressing what they already know about their environment and its capacity to deliver resources. This should be part of the process by which a community development plan is prepared.

11.7.3 Promoting conservation

Neither SMAP nor WESAMAR were designed specifically to protect biodiversity, but were intended to promote improved agriculture and rural development. Their open planning approach, however, can readily be adapted to a project area which contains a nature reserve. One way to do this is by including a project-specific system for environmental impact assessment (EIA).

"PARTICIPATIVE" DEVELOPMENT

The project design as proposed by experts

The project design as specified by Government

The project following a feasibility study by a Consultative Committee

The project after the final acceptance of a compromise plan

The project in implementation

What the people really wanted

There is no best way to perform EIA, and the design of the process in each case should fit the need in each case and the time and resources available (Carpenter & Maragos, 1989). It should reflect the fact that the EIA process is mainly one of thinking things through, rather than of only filling in a matrix or ticking off a checklist, and it therefore requires reasoning, understanding, wisdom, and synthesizing skills.

A project's EIA process should thus aim to ensure that no support is given to micro-projects which are likely to damage important features of the environment, such as nature reserves, mangroves, coral reefs, watersheds, estuaries, or rivers in the project area (Caldecott, 1994b). The first stage is for micro-project concepts to be discussed openly to see if they make sense, a process known as community screening and scoping (CSS). The aim of this is to talk and think through the likely effects of implementing each micro-project on the environment. The CSS process should lead to a decision on whether the effect of each micro-project on each key ecosystem will be beneficial, neutral, harmful, or unknown. If it is judged likely to be beneficial or neutral, the next stage is a feasibility study, which will allow the community and the project to decide whether or not to proceed further. This study will describe the environmental and social context of the micro-project, as well as defining an operational plan, a business plan, and a financing proposal.

If the micro-project is judged likely to be harmful the idea should be abandoned, while if its effects are unknown an environmental impact statement will be needed. The latter will give an opportunity to consider the ecological context of the proposed micro-project, and again to decide whether its effects on key ecosystems are likely to be beneficial, neutral, harmful, or unknown. The project's EIA process thus aims to help a community decide if it is safe to proceed, in terms of avoiding obvious kinds of damage to selected environmental resources. It should therefore complement rather than conflict with the national EIA laws applying to the project area, while allowing issues to be considered at a smaller-scale, more community-oriented level.

A project may need a more active approach to conservation than the EIA process alone, since local people may neither know how nor be inclined to accommodate the needs of wild species and natural ecosystems. Where urgent conservation problems exist, a project should be able to promote conservation among local people more actively than is allowed for by the model in Box

Fig. 11.3 The parable of the swing

11.6. Full-time staff with special training might be used to help community development plans reflect conservation priorities. These staff could be modelled on the village development promoters of the EU-funded North-East Arid Zone Development Project (NEAZDP) in Nigeria (W. Knight, personal communication, 1991), and their roles may also overlap with those of the Costa Rican parataxonomists (Chapter 6). Part-time staff may also be needed, whose role may be modelled on the village liaison assistants in the Cross River National Park Project in Nigeria (Chapter 3). These would remain in their own communities, where they would reinforce the educational message of the project, while aiding two-way communication between the project and local people.

These additions to the open planning model will increase labour needs while also creating a longer-term relationship between project and people than is described in Box 11.6. This may not be feasible for a project with a small budget or a limited lifetime, but may be essential if conservation is to be achieved. Donors should thus be encouraged to consider making longer-term investments in conservation projects than in other kinds of rural development project. These models also imply that a significant portion of the budget cannot be specified in advance of implementation, since much of the fieldwork will respond to community development plans rather than to the plans of project designers. Donors should therefore also be encouraged to consider investing flexibly in people and processes, rather than in the supply of standard packages of goods and services.

11.8 Conclusions

This chapter has reviewed procedures which can be built into conservation projects, to help people make decisions about which they feel positive, and which are also ecologically sound. These procedures are those of 'conservation psychology', and they include education, public relations, information services, economic advocacy, conflict management, group facilitation, negotiation, and open planning. They are all important because, in the mental and social arenas, the ways in which decisions are reached and conflicts resolved are often more important than exactly what decisions are made or how resources are divided. The environmental arena is rather different, since here the nature of decisions and their consequences affect people's relationship with their resource base. A sustainable project must be both environmentally and socially durable, and should therefore correct problems in the environmental arena in ways which are consistent with events in the mental and social arenas. This is more feasible than it may seem, since few questions in conservation have only one correct answer at any given time. This is because communities can be flexible in how they choose to use their resources.

A coastal village in Indonesia, for example, can use its near-shore waters in various ways, and two options are for fishing or diving ecotourism (Chapter 10). These two uses have quite different implications for management, and may be associated with different levels of income, but both can be a benign use of resources. They do share two preconditions, the first of which is that local people must understand something of the marine ecosystems involved, and the second is that they must have the power to control the use of those waters. If they are to have such control, then they must have the motivation, the physical ability and the authority to exclude outsiders.

If these conditions are met, then the community can choose its own sustainable management strategy by whatever social means it chooses. Projects in the Cyclops and Cenderawasih areas of Irian Jaya (Chapter 7), and in Maqueda Bay in the Philippines (Chapter 10), show that motivation to exclude others from fishing areas is easy to generate, provided that patrol boats are available and that the authority to patrol has been granted by government. A lack of local education, capacity, and authority are thus the practical constraints on tenure which conservation projects should first seek to overcome. Once they have done so, the creativity and adaptability of people can then help solve those conservation problems which continue to arise in the project area.

12

Summary and conclusions

12.1 The project concept

Once a group has decided to assess a potential conservation project area, the answers to certain questions will help to show if it is worthwhile to proceed further, and if so in what general direction (Chapter 8). They can be phrased in the form of a key (Box 12.1), but all call for judgements which would be aided by biological or anthropological research. If this cannot be done, the project designer should seek advice from people who know something about conservation, who are familiar with the ecosystem concerned, and who have spent time among the local people. Such expertise should first be sought locally, in the schools, universities, NGOs, or government offices close the project area, and then through national or international NGO networks.

When applying the key in Box 12.1 to a potential project area, it is important to remember the main aim of project design, which is to develop a *locally appropriate* project, adapted to the ecological and social realities of the project area. This may mean, for example, that if a project area is very diverse at a large scale, it can be sub-divided into smaller units until a scale is reached at which the answers to questions in Box 12.1 seem useful. A legally designated nature reserve, for example, may have been partly settled or otherwise used by various communities. Such a diverse patchwork cannot be managed for all purpose as a single unit, and any uniform management régime will be limited to what all the stakeholders can agree to. Rather than using an approach based on the 'lowest common denominator' in such an area, it is better to solve local problems comprehensively in each locality, and then to federate the many small management units in some way.

This would require forums for communication and procedures for managing conflicts among the various units. All stakeholders can then be encouraged to subscribe to a common ideal, such as sustainable development, and a common approach to defending regional interests against others. Possible outcomes range from a large intact reserve with some traditional use, buffer, and support zones around the edge, to a patchwork of small biodiversity refuges set in a landscape of community forests, fishing areas, farmlands, and settlements. Themes common to all successful outcomes would include secure tenure of land and other resources throughout the area, general agreement on all the boundaries involved, forums to discuss problems, and lines of communication outside the area to request assistance.

Box 12.1 A key to select the main features of a conservation project (for terms marked with an *asterisk, see the Glossary, Index, and Chapters 9 & 10)

Question 1. Is the reserve large enough to support indefinitely all the species, populations and ecological processes it now contains?
NO: go to question 2.
YES: go to question 4.

Question 2. Are there contiguous or nearby natural habitat areas, or areas which might regenerate under protection or in response to rehabilitation?
NO: go to question 3.
YES: consider enlarging the reserve; consider creating *buffer zones; consider creating *special zones in which to encourage recovery; go to question 3.

*Question 3. Is the area specially important for some *component of biodiversity or because it provides an important *environmental service?*
NO: seek advice on special factors which may justify intervention, or abandon project.
YES: go to question 4.

*Question 4. Are the *components of biodiversity or *environmental services involved likely to be degraded by continued current use?*
NO: consider community management using *traditional use zones; go to question 5.
YES: consider total protection as a *core area or biodiversity refuge; consider creating *buffer zones; go to question 5.

Question 5. Have any parts of the area been damaged in the recent past?
NO: go to question 6.
YES: assign *recuperation zones; go to question 6.

*Question 6. Are any *components of biodiversity in the area being degraded or are they about to be degraded by new forms of use?*
NO: withdraw and monitor events, or plan development of research, surveillance and tourism facilities.
YES: go to question 7.

Question 7. In view of the nature and distribution of influence over the future of the area, is a local solution to problems affecting it feasible?
NO: go to question 13.
YES: go to question 8.

Question 8. Relative to the size of the area, are there many people living nearby?
NO: adopt low-key programme of surveillance and public relations, while monitoring potential threats; go to question 9.
YES: go to question 9.

Question 9. Is the area used by local people?
NO: consider total protection as a *core area or biodiversity refuge; monitor potential threats.
YES: consider using *traditional use zones; monitor compliance and impacts; go to question 10.

Question 10. Are the local people culturally homogeneous, well established, and peaceful?
NO: go to question 11.
YES: build community-involvement programmes using established social leadership and regulations; negotiate agreements to involve them in protecting the reserve.

Question 11. Do the local people recognize a common leadership and set of values?
NO: go to question 12.
YES: strengthen emerging leadership and consensus; seek to encourage community feelings and involvement; identify groups with which to negotiate agreements to help manage the reserve.

Question 12. Can sufficient resources be deployed to the area to achieve compliance with conservation objectives?
NO: seek advice on special circumstances that may apply or special procedures that may be activated, or abandon project.
YES: plan intervention; secure resources; promote order; compensate, train, educate, and use group-facilitation techniques as necessary to begin building communities with which to work.

Question 13. Can you document the problem?
NO: seek advice on how to overcome political or other special constraints on access to information, or abandon project.
YES: activate existing procedures for spatial planning, environmental impact management or political appeal, if necessary lobbying for a change in policy.

12.2 From concept to implementation

In beginning a project, it is important that key participants understand its principles, that they agree with its strategy, and that they are prepared to support it being established. The identity of the participants who are seen as critical will depend on the scale of the project and the kinds of group or agency involved in it. A local NGO which hopes to spend US$1000 will require support at a different level than a multilateral development bank wishing to lend US$40 million, while the needs of international NGOs and bilateral donor agencies will fall somewhere between these extremes. Other sources of variation include differences in decision-making procedure among agencies and countries, and the fact that the needs for support of any project can be expected to change over time as the project evolves.

There follows a summary of the main sequence of events by which a project might be established. This is based partly on experience gained in the projects described in Chapters 2–7, and partly on Marty Fujita's work for The Nature Conservancy (TNC) in Indonesia during and after 1992, in which projects were established in the Lore Lindu and Morowali areas of Sulawesi (Cochrane, 1992; Schweithelm *et al.*, 1992). The steps given thus reflect the needs of international NGOs in establishing themselves in new project areas and countries, and in working with host governments and institutional donors. Future projects should adapt them in responding to different actors, circumstances, and priorities.

Step 1: *Origination.* The target reserve is selected from amongst possible alternatives, and the threats to it are assessed. The project area is visited, and ecological, economic, and social linkages between the reserve and its surroundings are studied to establish the nature of problems and opportunities to solve them. A draft concept paper is prepared, explaining the purpose of the project and the kinds of actions which it might undertake.

Step 2: *Informal consultation.* The concept is discussed informally with local informants, to identify perceived problems, constraints, interest groups, and conflicts that might arise among them. It is also discussed with government informants and with people in the donor community, to gauge the support which might be mobilized and to identify special concerns and interests which should be incorporated to increase support.

Step 3: *First central workshop.* Using the redrafted concept paper and verbal discussions, resources are obtained with which to hold a meeting of people whose endorsement is crucial to further progress. This will especially involve government people, since the aim of this workshop is primarily to obtain official approval and the support of an agency capable of sponsoring the next phase.

Step 4: *Feasibility study proposal.* The project may be conceived as necessarily complex and long term, and its sponsors may require a detailed feasibility study before making a commitment to support it in full. A detailed proposal, workplan and budget for such a study will thus often be needed (Chapter 8).

Step 5: *First local workshop*. As soon as the feasibility study is approved, representatives of local communities and other stakeholders (universities, businesses, etc.) who are likely to be affected by the project are invited to a meeting in or near the project area. The aim is to explain the current state of the project concept, the reasoning behind it, and the likely course of events in future, with the aim of confirming local support for the enterprise and preparing the way for later consultations. The kinds of benefit which local people might expect from conservation should be explained (Chapter 11), but they should be left in no doubt that many of those benefits will depend on their organizing themselves to protect their own interests, and to exploit the opportunities represented by the project.

Step 6: *First letter to local community leaders*. The first local workshop will have produced a consensus that the project is a good idea (**if it has not, return to step 2**). A short letter to confirm this understanding and to ask for co-operation with the feasibility study is sent to community leaders in the project area. The letter must be phrased in polite and positive terms, but should make no promises.

Step 7: *Community meetings*. At the beginning of the feasibility study, meetings are held to explain to local people directly what is about to happen and why, to introduce the strangers who will be working in the area, and to establish a friendly and supportive atmosphere for them.

Step 8: *Feasibility study*. The project is designed to the point where it can provide a clear statement of how to solve every significant problem which is perceived or predicted to affect the reserve, how much it will cost and, in general terms, who will do the work and when. The document is drafted, criticized informally by selected people, redrafted, discussed with people at the headquarters of the agency seeking to create the project, and then finalized and submitted to government for formal endorsement.

Step 9: *Second local workshop*. The main conclusions of the feasibility study are explained at a meeting of the same people who attended the first local workshop, with additions as necessary. The aim is to thank participants for their help and to explain what, if anything, is likely to happen next, and the expected timescale, which should be estimated very cautiously. Urgent conservation work which needs to be done meanwhile should be explained and arrangements made for it to be carried out.

Step 10: *Second letter to local community leaders.* Letters are written to the leaders of all communities which had contact with the feasibility study team, thanking their communities and themselves personally for their support. The likelihood of delay in implementing recommendations of the feasibility study should be emphasized. The letter may include an appeal to avoid killing certain animals or damaging certain habitats, or to collaborate with whatever conservation personnel remain in the project area pending full implementation of the project.

Step 11: *Second central workshop.* A meeting is held under the auspices of the official sponsor of the project, with the aim of formally presenting and explaining the results of the feasibility study to selected individuals and representatives of agencies whose support is needed to implement the project in full. It is important that everyone who feels that they should be invited to this workshop is in fact invited, and in good time to be able to attend if they wish to do so.

Step 12: *Processing the project proposal.* The feasibility study documents now comprise a detailed project proposal, which will be subjected to review, criticism, revision, and, perhaps, additional studies by the staff or consultants of donor agencies. As amended, it will finally be used as the basis for a financing proposal to the managing board of the donor agency concerned, who may make a budgetary allocation to support implementation.

Step 13: *Implementation.* A detailed workplan is developed, and roles, contracts, and management responsibilities are assigned in accordance with the procedures of the donor agencies and governments involved.

The implementation stage is likely to be reached between 1 and 5 years after step 1, by which time at least some of the international staff, government officials, and local people who were involved in designing the project will have moved on. Conditions in the project area may also have changed, and the conceptual framework and techniques to be used by the project may have become out of date. The scope for such change is shown by the evolution in project design over the period 1984–94 (Chapters 2–7). The implementing agency will therefore need to repeat the substance of several steps in order to ensure that the project still addresses the right issues and is still supported by all stakeholders.

12.3 From implementation to durable change

The steps followed in starting a project should ensure that all relevant groups and decision makers have the opportunity to contribute to its design and approval. The project design should have been guided by some of the practical suggestions in Chapter 8, and should reflect an appropriate choice amongst the options outlined in Chapters 9, 10, and 11. A summary of recommendations based on Chapters 8–11 is given in Box 12.2. These guidelines represent clues about what kind of project may be needed in any given location, and how to avoid some obvious mistakes while designing a project uniquely adapted to its own location. As the project moves into implementation, new circumstances will apply. It is essential to remember that, although a project must be designed to be implemented easily and well, the design is not itself the project. Instead, the design shows how the project will work, and it may establish the machinery through which it will work, but making it work is a different matter. This is where the procedures described in Chapter 11 become especially important, since the project will have to find ways to engage the interest of local people and to respond to their needs.

Box 12.2: Guidelines for designing conservation projects (from Chapters 8 to 11)

Part 1: Initial steps.

Guideline 1. Start the project inclusively. Make sure that key participants understand the principles of the project, that they agree with its strategy, and that they are prepared to support it being established.

Guideline 2. Design committees of stakeholders. Inclusive local committees will be needed to authorize action and expenditure, to ensure communication among stakeholders, and to invite assistance or to refer problems to task groups, forums, or to other levels of government if they cannot be solved locally.

Guideline 3. Design forums. Make sure that diverse interest groups can talk with each other regularly in order to build trust and negotiate peaceful solutions to conflicts of interest that may arise.

Guideline 4. Try to establish trust and maintain it. At every stage of a project, listen to stakeholders and hear them, show that you have heard, say what you will do, and do what you have said.

Guideline 5. Aim to achieve as much local participation as possible. Accept that this must be constrained by the interests of non-local communities and future generations.

Guideline 6. Employ and train local people. Local staff have many advantages over non-local ones, including, but not limited to, effective roles in long-term community development programmes in their own communities.

Guideline 7. Start small and local. Help to build local capacity and participation, by starting with small, successful pilot activities which can then be built upon.

Part 2: Education and participation.

Guideline 8. Combine environmental education with open planning. Guide each community towards being able to express their concept of the local environment, and the options, priorities, and need for assistance in improving the use of that environment for their own benefit.

Guideline 9. Use group facilitation techniques. Open planning should promote social cohesion, co-operation and consensus within the target community.

Guideline 10. Remember that education is powerful but can have side-effects. Consider the likely impact of new knowledge, thinking skills, and expectations on the people of the project area.

Guideline 11. Use education to help people understand their own needs. Promote the capacity to understand the value and vulnerability of environments, and to distinguish local needs from wants created by others.

Guideline 12. Promote awareness of why natural ecosystems are important. Adapt the message to local circumstances, for example by stressing their value in protecting soils, water supplies, and fisheries, and as sources of diversity to protect farming systems from crop disease and bad weather.

Part 3: Understanding the system.

Guideline 13. Combine saving with studying. Studies will show both how to use the components of biodiversity sustainably and how to teach about them interestingly. Use and interest will help people to continue saving them.

Guideline 14. Understand the needs of the constituents. Collect and manage biodiversity data to satisfy a wide variety of uses and users, ranging from prospectors and tourists to scientists and educators.

Guideline 15. Protect intellectual and other property rights. Document ethnobiological knowledge and promote biodiversity prospecting while linking success of the latter directly to solving conservation problems.

Guideline 16. Recognize diversity within communities. Collect and use information according to gender, age and socio-economic position, and analyze the impacts of all project activities accordingly.

Part 4: Conservation and development.

Guideline 17. Manage whole ecological units. Do not try to manage biodiversity in isolation from its geographical context, use ecological boundaries such as watersheds, and consider prevailing marine currents and riverine inflows. Manage buffer zones as whole ecosystems, rather than trying to simplify them to favour one particular kind of production.

Guideline 18. Close down open-access systems of resource exploitation. Help

Designing conservation projects

local people to obtain the ecological understanding, the legal authority based on traditional and communal ownership and usage-rights, and the practical capacity to control access to harvesting areas. Help exclusive users to develop a management régime and to comply with it.

Guideline 19. Ensure adequate funds for protection. Make permanent and adequate allocations for boundary patrols, consistent enforcement, and public relations. Providing alternative sources of income will not on its own stop people from overharvesting resources.

Guideline 20. Focus welfare programmes on special needs around the reserve. Always exchange assistance for compliance with conservation aims, through formal and monitored agreements.

Guideline 21. Use conservation biology. Understand why a species is declining before taking action. Manage habitat patchworks to maintain as much natural habitat as possible, including broad connected strips along which native species can disperse. If possible, keep duplicate populations in separate reserves, and plan for catastrophes affecting any one reserve.

Guideline 22. Use off-site action with caution. Ensure that off-site (*ex-situ*) and on-site (*in-situ*) activities are complementary, and do not compete for scarce conservation resources.

Guideline 23. Monitor wildlife harvests and adapt management systems. There should be the flexibility to respond quickly to changes in hunting and fishing technology and trade patterns.

Guideline 24. Identify and persuade local people to protect key habitats. Reserves for watersheds, mangroves, coral reefs, etc., are fundamental to the viability of local management régimes.

Guideline 25. Be aware of the risks of making resource use more efficient. Dangers include replacement of natural systems with artificial ones, and promoting local population growth and immigration.

Part 5: Finance and policy.

Guideline 26. Aim for balanced financing. Ensure that the project is funded and managed as a whole, so that the various components do not proceed at different speeds.

Guideline 27. Aim for durable financing. Ensure that all the funding options are considered, including local cost recovery, endowments, and other means to reduce dependency on external subsidies.

Guideline 28. Invest in tenure security and environmental awareness. These will help people to feel confident, involved, and interested in the project, and will reduce other costs.

Guideline 29. Present a convincing economic analysis. Be alert to the chance of documenting the economic role of ecosystems. Make sure that all costs and benefits of the project are taken into account, including those which are not easily quantifiable and especially those involving the risk of irreversible change.

Guideline 30. Encourage donor agencies to invest correctly. Long-term, flexible investments in people and processes are better than the supply of standardized packages of goods and services.

Guideline 31. Help projects to be supported by policy. Design them in the context of comprehensive national biodiversity management strategies.

Part 6: General points.

Guideline 32. Be prepared to take risks. It is not always possible to know in advance what will work and what will not, and conservationists should be willing to attempt extraordinary things.

Guideline 33. Remember that conservation is not self-contained. It is not only about choosing nature reserves, or managing wildlife populations and their habitats. Instead, it must involve changing people's minds, winning their support, and helping them to feel positive about the whole process.

Guideline 34. Promote feelings of confidence in the future. People cannot help if they are frightened or confused, and a project should help people feel more confident in a future containing high levels of biodiversity, than in any other.

Guideline 35. Use a gentle touch. Conservation problems are usually complex and subtle, and conservation projects require an approach closer to that of a diamond cutter, healer, or gardener, than that of a construction engineer.

The procedures which have been described aim to ensure that all parts of a project, as well as the whole, are both socially and environmentally durable. Projects designed in this way will emphasize finding practical ways for local people to be involved, while meeting the needs of governments and donor agencies. They will often be slow to show dramatic results, but will seek to make changes locally relevant and permanent by relying on local participation. Above all, they will try to ensure that the project can put down mental, social, financial, and institutional 'roots' among the people of the project area (Box 12.3).

Box 12.3 The 'root system' of a durable project

Mental roots begin with people becoming increasingly aware of the reasons for the project and the rules which will need to be followed if it is to work. They develop by people acquiring specific skills through training and greater understanding of their total environment through education. They develop fully through the confident exercise of participation in all aspects of the project, and the sense of empowerment which this creates.

Social roots begin with participation and empowerment, and are developed further through communal activities, whether encouraged by the project or spontaneous. They develop further as the security of resource tenure is increased at a community, family, or individual level, which also strengthens confidence and, from this position of security, encourages innovation in the use of resources.

Financial roots begin with the sense of security, which allows investment, and develop as innovation leads to returns on new investments. Meanwhile, project costs are minimized because public support reduces the frequency at which conservation rules are challenged, and also encourages voluntary co-operation with the project. They develop as new revenues are generated locally, and as those revenues are captured locally, and reach their fullest extent as budgetary responsibility is eventually assumed locally.

Institutional roots reinforce the others by guaranteeing local financial authority, permitting endowments to be established, establishing local stakeholder committees to oversee the project and to resolve conflicts of interest, and by providing a way for local priorities and actions to be integrated with national policies, procedures, and external funding.

12.4 Confidence and action

Conservation is becoming easier to practice as its ideas and ideals spread and are accepted. Someone who speaks out for the right of future people to live in a diverse environment, or for open planning, territorial use-rights in fisheries or local equity in biodiversity prospecting, is more likely to find understanding and support than would once have been the case. Allies have multiplied in unexpected places, and much conservation work now involves helping people to find the right words to express a sympathy they already feel.

Encouraging signs can be found in many events of the 1990s. They include the growth of global policies on sustainable development (Chapter 11), of national policies which promote decentralization (Chapters 5, 6, & 7), spatial planning and EIA (Chapters 4, 7, & 11), and the closure of open-access fisheries (Chapter 10). They also include the increased willingness of donor agencies to support open planning as the main way to deliver assistance to communities (Chapter 11).

Computer-based technologies also give grounds for optimism (Chapters 6 & 7). Global positioning systems, for example, allow the boundaries of community, corporate, and government land claims to be located accurately (Sirait *et al.*, 1994), while spatial data can be stored and managed within geographic information systems. The two together can provide definitive maps for use in resolving disputes affecting nature reserves and the people living around them (Fox, 1989; Prasodjo, 1992). This can make it easier to recruit local people as protectors of reserves and as managers of renewable resources.

Meanwhile, it has become possible to manage biodiversity data in ways which meet the needs of users ranging from local people, teachers, and school-children to tourists, planners, and scientists (Chapters 6 & 8). Lastly, networks of computer systems can also help local NGOs and other groups to increase their access to information, and to mobilize help against threats to local communities (Chapters 8 & 11). All these factors provide good reasons for conservationists to feel confident.

Many new attitudes and technologies, however, date only from the early 1990s. They will need to be proven and applied in real-life environments, where people and wild species actually live, and where ways must be found for them to coexist. No policy or technology can replace patient, small-scale work in thousands of real places, each one of which is different from all others. No project alone can do more than contribute to the further development of attitudes which previous projects have helped to create. Each project, however, can and should be made consistent with what conservationists wish the world to be, and by example help make it so. In this work, a good conservationist should always start from the local realities of the environment and its people, should have a clear idea of where the people can go to meet their own needs and those of wild species, and should accompany them on the journey for as long as local people provide a welcome.

Glossary

aboriginal (or autochthonous) people: all people who are indigenous to an area and who are descended from many generations of people who were also indigenous to it.

ADB: Asian Development Bank, a multilateral agency based in Manila with membership comprising most developing countries in the South-west, South, East and South-east Asian and Western Pacific regions, most Member States of the European Union, and also Australia, Canada, Japan, New Zealand, Switzerland, and the USA.

agroforestry: a general term for all farming systems which mix trees and woody shrubs with other crop plants and/or with livestock husbandry.

AMDAL: the Indonesian Environmental Impact Analysis or EIA process (*Analisis Mengenai Dampak Lingkungan*).

artificial ecosystem: one dominated by non-native species, by human buildings or clearances, or by native species which people have encouraged.

BAPEDAL: Indonesian Environmental Impact Management Agency (*Badan Pengendalian Dampak Lingkungan*).

BAPPEDA: Indonesian Regional Development Planning Agency (*Badan Perencanaan Pembangunan Daerah*).

BAPPENAS: Indonesian National Development Planning Agency (*Badan Perencanaan Pembangunan Nasional*).

bequest value: that part of the economic value of a species or ecosystem which reflects the fact that people want to retain the option for future generations to use it.

biodiversity (or biological diversity): the variety of distinct ecosystems within a living system, and/or the number of species within those ecosystems, and/or the range of genetic diversity within the populations of each of those species, thus meaning the richness and variety of living things in the world as a whole or in any place within it.

BIPHUT: Bureau of Forest Inventory and Mapping of the Indonesian Ministry of Forestry (*Biro Inventarisasi dan Perpetaan Hutan*).

buffer zone: an area of mainly natural ecosystems, outside but adjacent to a nature reserve, where human access and land-use are monitored or modified in support of the management of the reserve itself.

centre of plant diversity: one of about 250 locations identified by WWF & IUCN (1995) as places rich in native and endemic plant species, and which would collectively safeguard most of the world's wild plants, if all were protected.

CITES: Convention on International Trade in Endangered Species of Wild Fauna and Flora.

community: a group of people who perceive themselves to have a common cultural heritage and affinity to place, and who interact routinely in their daily lives.

Community Development Plan (CDP): a plan prepared by a community to describe its structure, its environment, and its development priorities, and to provide a context for proposals by which members of the community hope to meet their needs.

components of biodiversity: an expression from the Convention on Biological Diversity which refers collectively to ecosystems and habitats, species and communities, and genetic lineages; in particular (according to Annex I of the Convention), those which merit conservation concern because they are:

- 'Ecosystems and habitats: containing high diversity, large numbers of endemic or threatened species, or wilderness; required by migratory species; of social, economic, cultural or scientific importance; or, which are representative, unique or associated with key evolutionary or other biological processes.'
- 'Species and communities which are: threatened; wild relatives of domesticated or cultivated species; of medicinal, agricultural or other economic value; or social, scientific or cultural importance; or importance for research into the conservation and sustainable use of biological diversity, such as indicator species.'
- 'Described genomes and genes of social, scientific or economic importance.'

conservation: maintaining, restoring, or preventing damage to the components of biodiversity.

conservationist: one who sees promoting conservation as an important professional or personal task and who acts accordingly.

conservation project: a planned undertaking to conserve the components of biodiversity within a nature reserve.

conservation project area: the area in which a conservation project seeks to exert influence, normally including the whole ecological and economic region within which the target nature reserve is set.

core area (or biodiversity refuge): that part of a nature reserve which contains its native ecosystems in an intact state, and which is allocated to the protection and study of the components of biodiversity, to environmental monitoring, and to other forms of low-impact use. A core area may be divided into exclusion zones (no access); research zones (access by research and protection staff only); wilderness zones (for low-impact tourism); and limited use zones (for other restricted access).

debt-for-nature swap: an arrangement in which an instrument of indebtedness is obtained from the creditor of a third party by a conservation group, and is then given to the debtor in exchange for support for conservation.

direct use value: that part of the economic value of a species or ecosystem which reflects the fact that people use it or interact with it directly, whether for commercial, non-commercial or subsistence purposes, and whether for consumption (e.g. hunting, fishing, harvesting) or otherwise (e.g. recreation, research, education).

discount rate: the rate (as percentage of current value) by which the economic value of future costs or benefits is discounted relative to the economic value of current costs or benefits.

discounting: the practice of adjusting the economic value of costs and benefits according to when they are incurred or received, thus reflecting the assumption that benefits gained or lost are worth more if this happens sooner rather than later.

economic development: the process of meeting those human needs which can be given an economic value more completely than they were before.

economic valuation: the process of estimating the economic value of resources, processes, goods, and services, usually expressed in monetary units.

economic valuation techniques: ways to estimate economic value, using data from real markets (e.g. effects on production, loss of earnings, preventative expenditure, and replacement costs), from imaginary markets (e.g. willingness to pay for gain or to be paid for loss), or from indirect markets (e.g. trade in indicators or equivalents of environmental services, including property values, travel costs, and product substitution).

economic value: the sum of anything's direct and indirect use values, option and bequest values, and existence values.

ecoregion (or ecological region): an ecologically distinct biogeographical unit comprising part of a country or a transfrontier area.

ecosystem: a community of organisms considered as a whole with their physical environment.

ecotourism: a variety of nature-based tourism which uses but does not damage natural ecosystems, and which contributes both to their continued protection and to the well-being of local people.

endemic: a species or higher taxon which occurs in the wild nowhere other than in a particular place, and which has not done so in the historical past.

endemic bird area: one of about 221 places identified by ICBP (1992, 1995) where there is overlap among two or more birds, each of which have total breeding ranges of $50\,000\ \text{km}^2$ or less. They together amount to less than 5% of the world's land area, but about 26% of all bird species are confined to them.

environmental education: the process of teaching and learning about the significance of the environment for all human activities, and about how to appreciate and use the environment sustainably.

environmental impact assessment (or analysis, EIA): a set of procedures to help people think through and anticipate the likely impacts of a project on other people and on the environment, and to define ways to avoid, mitigate, or compensate for them.

environmental service: the function of an ecosystem which gives rise to its indirect use value, for example its capacity to safeguard water supply, to prevent floods and erosion, to store carbon, or to maintain the productivity of fish or wildlife populations.

ethnobiology: the study of the role of organisms in people's culture, myths, languages, medical practices, etc.

European Commission (EC, or Commission of the European Communities, CEC): the European Union's technical arm and civil service, based in Brussels and with Delegations or Representations in other countries.

European Union (EU): an association of 15 Member States, comprising Austria, Belgium, Denmark, Eire, Finland, France, Germany, Greece, Italy, Luxembourg, the Netherlands, Portugal, Spain, Sweden, and the United Kingdom.

existence value: that part of the economic value of a species or ecosystem which reflects the fact that people want it preserved for its own sake without any intention of using it.

exotic: a species or higher taxon which has been introduced by people to an area, or which has colonised the area in the historical past.

ex-situ **(or off-site):** located outside the site of origin (e.g. a wild organism living in a zoo or botanic garden).

forest-dependent people: those who obtain an essential part of their means of subsistence by harvesting organisms within natural forests.

forest-dwelling people: those who are forest-dependent and who normally sleep at night in temporary structures beneath the canopy of a natural forest.

GDP: gross domestic product, an estimate of total economic activity within a specified territory, calculated mainly with reference to industrial output and transactions in markets for goods and services.

GIS (Geographical Information System): a tool for storing and updating spatial information from paper maps and remote imagery using computer-based techniques for encoding, analyzing, and displaying multiple data layers (Burroughs, 1986).

GPS (Global Positioning System): a tool for locating positions accurately by triangulating from multiple satellites in high Earth orbit.

habitat: the locality or ecosystem type in which an organism lives in the wild state.

habitat area: a region containing natural ecosystems within which native species can be expected to occur.

human development: the process of meeting human needs more completely than they were before.

human needs: the social, bodily, intellectual, spiritual, and emotional needs of people.

ICDP: integrated conservation and development project.

INBio: National Biodiversity Institute of Costa Rica (*Instituto Nacional de Biodiversidad*).

indigenous (or native) people: all people who were born within the area concerned.

indigenous (or native) species: all species which arose in the area concerned by local evolution, or which arrived in that area without the aid of people.

indirect use value: that part of the economic value of a species or ecosystem which reflects its contribution to sustaining economic activity and protecting property.

infrastructure zone: that part of a nature reserve which is needed for infrastructure or intensive use in support of administration, maintenance, public services, and control programmes in accordance with the area's management plan.

in-situ **(or on-site):** located within the site of origin (e.g. a wild organism living in a nature reserve).

institutional failure: the inability of an institution to control its environmental impacts adequately, due to insufficient resources or inadequate public accountability.

intensive use zone: that part of a conservation area which is needed for trail networks, camp-sites, and viewing points to service public needs in accordance with the area's management plan.

IUCN Category I Protected Areas: scientific reserves and wilderness areas, which are defined as being largely free of human disturbance, and are available mainly for scientific research, environmental monitoring, and non-mechanized, non-disruptive forms of tourism.

IUCN Category II Protected Areas: national parks and equivalent reserves, which are defined as being large natural areas managed by national or nationally recognized authorities to protect the ecological integrity of ecosystems.

IUCN Category III Protected Areas: natural monuments, which are defined as being natural or natural–cultural features of outstanding interest, and are managed so as to protect those features.

IUCN Category IV Protected Areas: habitat and wildlife management areas, which are defined as being subject to human intervention to maintain ecological conditions and populations of certain species (e.g. for harvesting or to safeguard migrants).

IUCN Category V Protected Areas: protected landscapes and seascapes, which are defined as being the product of harmonious interaction between people and nature, often combining scenic beauty with traditional patterns of human settlement and resource use.

keystone species: a species which if lost from an ecosystem will cause the loss of other species from the same ecosystem.

KLH: Indonesian Ministry of State for Population and Environment (*Kependudukan dan Lingkungan Hidup*). Since 1993, divided into the Ministry of State for Population and the Ministry of State for Environment.

KSDA: Natural Resources Office (*Kantor Sumber Daya Alam*), the technical arm of the Indonesian PHPA.

LIPI: Indonesian Institute of Sciences (*Lembaga Ilmu Pengetahuan Indonesia*).

local people: those who were born or who have settled in an area, and who normally live there.

management plan: a plan for managing a nature reserve by specifying appropriate boundaries, protective actions, use zones, roles, and responsibilities within the management service, visitor use patterns, and programmes for monitoring, public relations, education and training.

market failure: an arrangement in which the people who benefit from damaging environments do not also bear the full cost of that damage.

matrix environment: any majority background state, usually an artificial ecosystem, within which fragments of natural ecosystems can be distinguished.

MIRENEM: the Costa Rican Ministry of Natural Resources, Energy and Mines (*Ministerio de Recursos Naturales, Energía y Minas*).

natural ecosystem: one dominated by native species whose diversity, distribution, and relative abundance is similar to that expected in the absence of human interference.

nature reserve: any viable sample of a natural ecosystem which is, or might be, or should be, set aside by law or custom mainly for conservation purposes. The term owes much to the Costa Rican *Area de Conservación*, but to translate this directly would have invited confusion with the use of 'conservation area' in the Pacific, where it means a settled, mainly agricultural landscape, managed by local landowners (SPREP, 1992), which itself overlaps with the IUCN Category V Protected Area (CNPPA, 1990). Nature reserves are broadly equivalent to Protected Areas in IUCN Categories I and II, some of those in Categories III and IV, and parts of those in Category V.

NGO: non-governmental organization, including private voluntary, non-profit, grass-roots, and community self-help groups.

open-access exploitation: an arrangement in which resources are exploited competitively by different groups and individuals.

open planning: a process of planning a project in which people are encouraged and enabled to become involved at every stage from the initial concept to the final evaluation.

option value: that part of the economic value of a species or ecosystem which reflects the fact that people want to retain their option to use it in future.

paper park: a nature reserve which exists only on paper, a status retained until its management plan is implemented.

PHPA: Directorate-General of Forest Protection and Nature Conservation (*Perlindungan Hutan dan Pelestarian Alam*) of the Indonesian Ministry of Forestry.

planning failure: the allocation of ecosystems to uses which they cannot sustain because of their inherent fragility.

policy failure: the persistence of government policies which encourage the over-harvesting of renewable resources, the release of wastes at a rate greater than the ability of ecosystems to absorb them, or otherwise fail to take adequate account of environmental concerns.

project plan: an account of what will be done with existing resources to achieve specific objectives.

project proposal: an account of what would be done with new or additional resources if they were to be obtained.

recuperation zone: that part of a nature reserve which is being allowed to recover indefinitely from disturbance.

RePPProT: Regional Physical Planning Programme for Transmigration, a national mapping exercise undertaken in the 1980s with British Government support for the Indonesian Ministry of Transmigration.

SCUBA: self-contained underwater breathing apparatus.

shifting cultivation (*swidden, kaingin, ladang,* etc.): a form of agriculture in which fields are cleared by hand and fire, and cropped discontinuously. In established systems, short periods of cropping alternate with longer periods of fallow prior to reuse of the same land, while in pioneer systems significant amounts of climax vegetation are cleared each year.

SINAC: the Costa Rican National System of Conservation Areas (*Sistema Nacional de Areas de Conservación*).

SNR: Indonesian strict nature reserve (*Cagar Alam*).

SMAP: Southern Mindanao Agricultural Programme (PO Box 8333, Davao City, Philippines).

social (or community) forestry: any forest management process which closely involves local people in managing and directly benefiting from forestry and tree-growing activities.

sovereign debt: a debt owed by one government to another.

stakeholder: someone with an interest in what happens to resources in a project area, or someone who may gain or lose something in a dispute over resources.

support zone: an area embracing those communities whose active support is needed by a project, and who may therefore be allocated special help or guidance during the project.

sustainable (or durable) development: a form of development which is likely to allow human needs to be met in the future at least as well as they are in the present.

sustainable use of wild species: a use which does not reduce the potential of the target species or any other species to be used, and which does not impair its viability or that of any other species, and also which does not impair the viability of any supporting and dependent ecosystem.

tenure: the holding by people of the means to satisfy their needs and to determine their future.

TGHK: Indonesian consensus classification of forest function (*Tata Guna Hutan Kesepakatan*).

traditional people: those whose social life and access to natural resources is acknowledged by themselves to be mainly governed by customary procedures, rather than by national laws.

traditional use zone: that part of a nature reserve in which exclusive privileges to continue using certain resources in traditional ways have been allocated to specified communities, in exchange for their assistance in managing the reserve as a whole.

TURF: territorial use-rights in fisheries.

upstream analysis: the study of threats to a nature reserve which do not arise locally and cannot necessarily be solved by local action.

WESAMAR: Western Samar Agricultural Resources Development Programme (ATI-Regional Fishermen's Training Centre, 6700 Catbalogan, Samar, Philippines).

Bibliography

Abel, N. O. J., Drinkwater, M. J., Ingram, J., Okafor J. C., & Prinsley, R. T. (1989a) *Guidelines for Training in Rapid Appraisal for Agroforestry Research and Extension.* Commonwealth Science Council, Commonwealth Secretariat (London, UK).

Abel, N. O. J., Drinkwater, M. J., Ingram J., Okafor J. C., & Prinsley, R. T. (1989b) *Agroforestry for Shurugwi, Zimbabwe: Report of an Appraisal Exercise for Agroforestry Research and Extension.* CSC Agriculture Programme, Agroforestry, series number CSC (89) AGR-14, Technical Paper 277. Commonwealth Secretariat (London, UK).

Abou Bakr Ahmed Ba Kader, Abdul Latif Tawfik El Shirazy Al Sabbagh, Mohamed Al Sayyed Al Glenid, & Mouel Yousef Samarrai Izzidien (1983/ 1403). *Basic Paper on the Islamic Principles for the Conservation of the Natural Environment.* IUCN Environmental Policy & Law Paper 20. World Conservation Union (IUCN, Gland, Switzerland).

Abraham, E. R. G., Sargento, J. O., & Torres, C. S. (1992) *Census and On-Ground Verification of Farms in the Mt Makiling Forest Reserve: Final Report.* College of Forestry & Environment and Resource Management Project, University of the Philippines at Los Baños (UPLB, College, the Philippines).

Adams, W. M. (1987) Approaches to water resource development, Sokoto Valley, Nigeria: the problem of sustainability. Pp. 307–25 in *Conservation in Africa: People, Policies and Practice* (edited by D. Anderson & R. Grove). Cambridge University Press (Cambridge, UK).

Adams, W. M. & Thomas, D. H. L. (1993) Mainstream sustainable development: the challenge of putting theory into practice. *Journal of International Development,* 5: 591–604.

ADB (1992a) *Conservation of Tropical Forest Ecosystems and Biodiversity: Final Report.* Feasibility Study prepared under TA No. 1430 (INO/Indonesia) by Deutsche Forst-Consult. Asian Development Bank (ADB, Manila, Philippines).

ADB (1992b) *Biodiversity Conservation Project in Siberut and Ruteng in Indonesia.* Pre-appraisal prepared under TA No. 1430 (INO/Indonesia) by S. T. Qadri, J. O. Caldecott, & R. A. Kramer. Asian Development Bank (ADB, Manila, Philippines).

ADB (1992c) *Environment and Natural Resources Masterplan for Hainan Province, China.* Prepared under TA No. 1464 (PRC/People's Republic of

271

China) by Environmental Resources Ltd. Asian Development Bank (ADB, Manila, Philippines).

ADB (1992d) *Action Plan for Environmental Improvement and Management of Hainan Province, China.* Prepared under TA No. 1464 (PRC/People's Republic of China) by Environmental Resources Ltd. Asian Development Bank (ADB, Manila, Philippines).

ADB (1994) *Progress of the Biodiversity Conservation Project in Indonesia: Report of a review mission in March 1994.* Asian Development Bank (ADB, Manila, Philippines).

Aldhous, P. (1991) 'Hunting licence' for drugs. *Nature*, **353**: 290.

Allen, E. F. (1948) The bearded pig. *Malayan Nature Journal* **3**: 98–99.

Anderson, J. A. R. (1972) *A Note on the Plantations of Dipterocarpaceae at Hauerbentes, Java, Indonesia.* Mimeo.

Anderson, J. A. R. (1975) The potential of illipe nuts (*Shorea* spp.) as an agricultural crop. *Proceedings of the FAO, BIOTROP & LBN-LIPI Symposium on South-east Asian Plant Genetic Resources, 20–22 March 1975* (Bogor, Indonesia).

Angel, M. V. (1992) Managing biodiversity in the oceans. Pp. 23–62 in *Diversity of Oceanic Life: an Evaluative Review* (edited by M. N. A. Peterson). Center for Strategic and International Studies (Washington, DC, USA).

Anonymous (undated) *Understanding the User's Perspective in Development: a Parable to Remember.* Attributed to Robert Rhoades. User's Perspective with Agricultural Research and Development (UPWARD).

Anonymous (1994) RI log output to be cut. *Jakarta Post*, 15th June 1994.

Anthonio, Q. B. O. (1990) *Observations on Agricultural Economics in the Support Zone.* Appendix 6 of *Cross River National Park, Okwangwo Division: Plan for Developing the Park and Its Support Zone.* World Wide Fund for Nature (WWF, Godalming, UK).

Applegate, G. B., Chamberlain, J., Daruhi, G., Feigelson, J. L., Hamilton, L., McKinnell, F. H., Neil, P. E., Rai, S. N., Rodehn, B., Statham, P. C., & Stemmermann, L. (1990) Sandalwood in the Pacific: a state-of-knowledge synthesis and summary from the April 1990 Symposium. Pp. 1–11 in *USDA Forest Service General Technical Report PSW-122.* United States Department of Agriculture (Washington, DC, USA).

Areola, O. (1987) The political reality of conservation in Nigeria. Pp. 277–92 in *Conservation in Africa: People, Policies and Practice* (edited by D. Anderson & R. Grove). Cambridge University Press (Cambridge, UK).

ASC (1993) *An Information Model for Biological Collections (March 1993 Version).* ASCII Text via TAXACOM FTP Server at huh.harvard.edu (/pub/standards/asc/ascmodel.txt) and figures in Postscript form (/pub/standards/asc/ascfig1.ps and /ascfig2.ps). Association of Systematics Collections (Washington, DC, USA).

Ashton-Jones, N. J. (1992a) *Oban Hills Integrated Rainforest Conservation Project: Project Manager's Report to ODA for the Period April 1991 to March 1992 – ODA Project No. 77 (WWF No. 3264).* World Wide Fund for Nature (WWF, Godalming, UK).

Ashton-Jones, N. J. (1992b) *Development of the Northern Sector of Cross River National Park: Project Manager's Report to ODA for the Period April 1991 to March 1992 – ODA Project (WWF No. 3916).* World Wide Fund for Nature (WWF, Godalming, UK).

Avé, W. & Sunito, S. (1990) *Medicinal Plants of Siberut.* World Wide Fund for Nature (WWF, Gland, Switzerland).

Balmford, A., Leader-Williams, N. & Green, M. J. B. (1995)   Parks or arks: where to conserve threatened mammals? *Biodiversity and Conservation*, **4**: 595–607.

Balmford, A., Mace, G. M., & Leader-Williams, N. (in press)   Redesigning the Ark: setting priorities for captive breeding. *Conservation Biology*.

BAPEDAL (1993)   *Government Regulation of the Republic of Indonesia Number 51 of 1993 Regarding Environmental Impact Assessment*. Non-binding translation published by the Environmental Impact Management Agency (BAPEDAL) & Environmental Management Development in Indonesia (EMDI) Project (Jakarta, Indonesia).

BAPEDAL (1994)   *A Guide to Environmental Impact Assessment in Indonesia*. Environmental Impact Management Agency (BAPEDAL) & Environmental Management Development in Indonesia (EMDI) Project (Jakarta, Indonesia).

BAPPENAS (1991)   *Biodiversity Action Plan for Indonesia*. National Development Planning Agency (BAPPENAS, Jakarta, Indonesia).

Barbier, E. B. & Aylward, B. (1992)   *What is Biodiversity Worth to a Developing Country? The Pharmaceutical Value of Biodiversity and Species Information*. London Environmental Economics Centre (London, UK).

Barbier, E. B., Adams, W. M., & Kimmage, K. (1991)   *Economic Valuation of Wetland Benefits: the Hadejia-Jama'are Floodplain, Nigeria*. World Conservation Union (IUCN, Gland, Switzerland).

Barrett, S. (1988)   *Economic Guidelines for the Conservation of Biological Diversity*. World Conservation Union (IUCN, Gland, Switzerland).

Barzetti, V. (editor, 1993)   *Parks and Progress: Protected Areas and Economic Development in Latin America and the Caribbean*. World Conservation Union (IUCN, Gland, Switzerland).

Beddington, J. (1995)   Fisheries: the primary requirements. *Nature*, **374**: 213–14.

Bennett, J. & Caldecott, J. O. (1988)   *Development of Model Rattan Nurseries at Long Laput and Long Tungan on the Baram River, Sarawak*. Land Associates International Ltd (London, UK) and Sarawak Forest Department (Kuching, Sarawak, Malaysia).

Bennett, E. L. & Caldecott, J. O. (1989)   Primates of Peninsular Malaysia. Pp. 355–63 in *Ecosystems of the World, 14B – Tropical Rain Forest Ecosystems: Biogeographical and Ecological Studies* (edited by H. Leith & M. J. A. Werger). Elsevier (Amsterdam, the Netherlands).

Bennett, E. L. & Reynolds, C. J. (1993)   The value of a mangrove area in Sarawak. *Biodiversity and Conservation* **2**: 359–75.

Bleakley, S. (1994) *Coastal Conflicts: Past Mistakes, Future Directions*. Master's Dissertation, University of Edinburgh (Edinburgh, UK).

Bleakley, S., Craig, E., Drew, M., Fooks, M., Garrad, S., Kanis, B., Lilley, S., Sidaway, R. (1994)   *Developing Community Involvement in the Green Belt: Untapped Potential Waiting for a Spark*. Institute of Ecology and Resources Management, University of Edinburgh (Edinburgh, UK).

Boado, E. L. (1988)   Incentive policies and forest use in the Philippines. Pp. 165–203 in *Public Policies and the Misuse of Forest Resources* (edited by R. Repetto & M. Gillis). Cambridge University Press (Cambridge, UK).

Bohnsack, J. A. (1994)   Marine reserves: they enhance fisheries, reduce conflicts, and protect resources. *NAGA, The ICLARM Quarterly*, July 1994: 4–7 (excerpted from *Oceanus*, Fall 1993, **36**).

Bohnsack, J. A. & Sutherland, D. L. (1985)   Artificial reef research: a review with recommendations for future priorities. *Bulletin of Marine Science*, **37**: 11–39.

Bonner, J. (1994)   Wildlife's roads to nowhere? *New Scientist*, 20 August 1994 (1939): 30–4.

Boo, E. (1990)   *Ecotourism: the Potentials and the Pitfalls*. World Wildlife Fund (Washington, DC, USA).

Boorman, J. & Roche, P. (1957–1961)   *The Nigerian Butterflies* (in six volumes). Ibadan University Press (Ibadan, Nigeria).

Bowles, M. L. & Whelan, C. J. (editors, 1994)   *Restoration of Endangered Species: Conceptual Issues, Planning, and Implementation*. (Cambridge University Press, Cambridge, UK).

Braatz, S., Davis, G., Shen, S., & Rees, C. (1992)   *Conserving Biological Diversity: A Strategy for Protected Areas in the Asia-Pacific Region*. World Bank Technical Paper No. 193. The World Bank (Washington, DC, USA).

Bradley Martin, E. (1983)   *Rhino Exploitation: the Trade in Rhino Products in India, Indonesia, Malaysia, Burma, Japan & South Korea*. World Wildlife Fund (WWF, Hong Kong, China).

Brandon, K. E. and Wells, M. (1992)   Planning for people and parks: design dilemmas. *World Development* **20**: 557–70.

Bridger, G. A. & Winpenny, J. T. (1987)   *Planning Development Projects: a Practical Guide to the Choice and Appraisal of Public Sector Investments, Second Edition*. Her Majesty's Stationery Office (London, UK).

Brillantes, A. B., Jr (1993)   *The Philippine Local Government Code of 1991: Issues and Concerns in the Environment Sector*. Report No. 5, Environment and Resources Management Project (College, Laguna, Philippines, & Halifax, Nova Scotia, Canada).

British Council (1994)   *Indonesia – ADB Institutional Strengthening of Biodiversity Conservation TA Study: Inception Report*. The British Council (Jakarta, Indonesia).

Brock, R. E. (1985)   Preliminary study of the feeding habits of pelagic fish around Hawaiian fish aggregation devices or: Can fish aggregation devices enhance local fisheries productivity? *Bulletin of Marine Science*, **37**: 114–28.

Brooke, M. (1913)   *My Life in Sarawak*. Oxford University Press (Oxford, UK).

Brosius, J. P. (1986)   River, forest and mountain: the Penan Gang landscape. *Sarawak Museum Journal*, **36**: 173–84.

Brosius, J. P. (1993)   Penan of Sarawak. Pp. 142–3 in *State of the Peoples: a Global Human Rights Report on Societies in Danger*. Published for Cultural Survival by Beacon Press (Boston, Massachusetts, USA).

Burkill, I. H. (1966)   *A Dictionary of the Economic Products of the Malay Peninsula*. Ministry of Agriculture and Co-operatives (Kuala Lumpur, Malaysia).

Burnie, D. (1994)   Ecotourists to paradise. *New Scientist*, 16 April 1994 (1921): 23–7.

Burroughs, P. A. (1986)   *Principles of Geographical Information Systems for Land Resources Assessment*. Clarendon (Oxford, UK).

Caldecott, J. O. (1980)   Habitat quality and populations of two sympatric gibbons (Hylobatidae) on a mountain in Malaya. *Folia Primatologica*, **33**: 291–309.

Caldecott, J. O. (1986)   *An Ecological and Behavioural Study of the Pig-tailed Macaque*. Contributions to Primatology, vol. 21: 1–259. Karger (Basel, Switzerland).

Caldecott, J. O. (1987a)   *Gaharu as a Potential Crop in Sarawak*. Mimeo: National Parks & Wildlife Office, Sarawak Forest Department (Kuching, Sarawak, Malaysia).

Caldecott, J. O. (1987b) *The Potential of Illipe Nut Cultivation in Sarawak.*
Mimeo: National Parks & Wildlife Office, Sarawak Forest Department
(Kuching, Sarawak, Malaysia).

Caldecott, J. O. (1987c) *Rattan Potential in Sarawak.* Mimeo: National Parks &
Wildlife Office, Sarawak Forest Department (Kuching, Sarawak, Malaysia).

Caldecott, J. O. (1987d) Medicine and the fate of tropical forests. *British Medical
Journal* **295**: 229–30.

Caldecott, J. O. (1987e) A leaf from nature's book. *The Dentist* **4**: 36–6.

Caldecott, J. O. (1987f) Earthlife expedition finds Sumatran rhino. *Earthlife
News*, No. 6: 23–4.

Caldecott, J. O. (1987g) *Rhinoceros Task Force: Briefing Document for the
Inaugural Meeting on 22 August 1987.* Sarawak Forest Department (Kuching,
Sarawak, Malaysia).

Caldecott, J. O. (1988a) *Hunting and Wildlife Management in Sarawak.* IUCN
Tropical Forest Programme Monographs, vol. 7: 1–168. World Conservation
Union (IUCN, Gland, Switzerland).

Caldecott, J. O. (1988b) Climbing towards extinction: a crisis for climbing palms.
New Scientist, 9 June 1988 (1616): 62–6.

Caldecott, J. O. (1988c) *A Commercial Natural Products Institute in Sarawak,
Malaysian Borneo: Proposal for a Feasibility Study.* Land Associates
International Ltd (London, UK).

Caldecott, J. O. (1988d) A variable management system for hill forests in
Sarawak. *Journal of Tropical Forest Science,* **1**: 103–13.

Caldecott, J. O. (1988e) *A Masterplan for the Sustainable Use of Lands, Forests
and Waters in the Upper Baram Area of Sarawak (Malaysian Borneo).* Land
Associates International Ltd (London, UK).

Caldecott, J. O. (1988f) *An Integrated Rainforest Conservation and Rural
Development Masterplan for the Oban Hills Region of Cross River State,
Nigeria.* World Wide Fund for Nature (WWF, Godalming, UK) and Nigerian
Conservation Foundation (NCF, Lagos, Nigeria).

Caldecott, J. O. (1991a) Eruptions and migrations of bearded pig populations.
Bongo (Berlin), **18**: 233–43.

Caldecott, J. O. (1991b) Monographie des Bartschweines (*Sus barbatus*). *Bongo
(Berlin),* **18**: 54–68.

Caldecott, J. O. (1991c) Ecology of the bearded pig in Sarawak. Pp. 117–29 in
Biology of Suidae (edited by R. H. Barrett & Spitz). IRGM/INRA (Toulouse,
France).

Caldecott, J. O. (1991d) *The Cross River National Park Project, Nigeria:
Implications for Project Design and Development.* East–West Centre
(Honolulu, Hawai'i, USA) and World Wide Fund for Nature (WWF,
Godalming, UK).

Caldecott, J. O. (1991e) *The Cross River National Park Project, Nigeria:
Operational Experience During the Start-up Phase.* East–West Centre
(Honolulu, Hawai'i, USA) and World Wide Fund for Nature (WWF,
Godalming, UK).

Caldecott, J. O. (1992a) Biodiversity management in Indonesia: a personal
overview of current conservation issues. *Tropical Biodiversity,* **1**: 57–62.

Caldecott, J. O. (1992b) *Concepts and Strategies in Designing Integrated
Conservation and Development Projects for Protecting Indonesian Biodiversity
Assets.* DHV Consultants (Bogor, Indonesia).

Caldecott, J. O. (1993a) *Opportunities for Biodiversity Management in the
Makiling Forest Reserve.* Report No. 10, Environment and Resources

Management Project (College, Laguna, Philippines & Halifax, Nova Scotia, Canada).

Caldecott, J. O. (1993b) *Conservation and Environmental Education in the Pacific South (Brunca) Region of Costa Rica.* Living Earth Foundation (London, UK).

Caldecott, J. O. (1994a) *Conservation of Biological and Cultural Diversity in Irian Jaya, Indonesia.* World Wide Fund for Nature (WWF, Jakarta, Indonesia).

Caldecott, J. O. (1994b) *Environmental Manual for the WESAMAR Program,* Western Samar Agricultural Resources Development Programme (WESAMAR, Catbalogan, Samar, Philippines).

Caldecott, J. O. & Alikodra, H. S. (1992) *Biodiversity Management for Sustainable Development in Costa Rica.* Ministry of State for Population and Environment (KLH, Jakarta) and School for Resource and Environmental Studies, Dalhousie University (Halifax, Nova Scotia, Canada).

Caldecott, J. O., Bennett, J. G., & Ruitenbeek, H. J. (1989) *Cross River National Park, Oban Division: Plan for Developing the Park and Its Support Zone.* World Wide Fund for Nature (WWF, Godalming, UK).

Caldecott, J. O., Blouch, R. A., & Macdonald, A. A. (1993) *The bearded pig (Sus barbatus).* Pp. 136–45 in *Pigs, Peccaries, and Hippos – Status Survey and Conservation Action Plan* (edited by W. L. R. Oliver). World Conservation Union (IUCN, Gland, Switzerland).

Caldecott, J. O. & Caldecott, S. (1985) A horde of pork. *New Scientist,* 15 August 1985 (1496): 32–5.

Caldecott, J. O. & Fameso, T. F. (1991) *TFAP Nigeria: Findings of the Preliminary Conservation and Environment Study Mission.* Commission of the European Communities (Brussels, Belgium).

Caldecott, J. O., Jenkins, M., Johnson, T., & Groombridge, B. (1994) *Priorities for Conserving Global Species Richness and Endemism.* World Conservation Press (Cambridge, UK).

Caldecott, J. O. & Labang, D. (1988) *Rhinoceros Surveys in Northern Sarawak.* National Parks and Wildlife Office, Sarawak Forest Department (Kuching, Sarawak, Malaysia).

Caldecott, J. O. & Mundy, J. R. J. H. (1987) *Forest Enterprise Zones.* Land Associates International Ltd (London, UK).

Caldecott, J. O., Oates, J. F., & Ruitenbeek, H. J. (1990) *Cross River National Park, Okwango Division: Plan for Developing the Park and Its Support Zone.* World Wide Fund for Nature (WWF, Godalming, UK).

Caldecott, J. O., Oates, J. F., Gadsby, E. L., & Edet, C. A. (1990) *Gorilla-based and Other Tourism Development Potential.* Appendix 4 of *Cross River National Park, Okwangwo Division: Plan for Developing the Park and Its Support Zone.* World Wide Fund for Nature (WWF, Godalming, UK).

Caldecott, J. O. & Ruitenbeek, H. J. (1989) *Dedicated Sovereign Debt Servicing for the Management of Cross River National Park, Nigeria, and its Support Zone.* World Wide Fund for Nature (WWF, Godalming, UK).

Callister, D. J. (1992) *Illegal Tropical Timber Trade: Asia-Pacific.* Trade Records Analysis for Flora and Fauna in Commerce (TRAFFIC, Cambridge, UK).

Carpenter, R. A. & Maragos, J. E. (1989) *How to Assess Environmental Impacts on Tropical Islands and Coastal Areas.* Environment and Policy Institute, East–West Centre (Honolulu, Hawai'i, USA).

Carruthers, I. & Chambers, R. (1981) Rapid appraisal for rural development. *Agricultural Administration,* 8(6): 407–22.

Cartwright, J. R. & Frame, G. W. (1986)   What is or should be a "buffer zone"? Pp. 32–5 in *Proceedings of the IUCN/TFAG Meeting on the Management of Rainforest Buffer Zones, Merida (Venezuela), 29 Nov–2 Dec 1986.* World Conservation Union (IUCN, Gland, Switzerland).

Castillo-Muñoz, R. (1983)   Geology. Pp. 47–62 in *Costa Rican Natural History* (edited by D. H. Janzen). University of Chicago Press (Chicago, Illinois, USA).

Caughley, G. (1977)   *Analysis of Vertebrate Populations.* Wiley (Chichester, UK).

Caughley, G. (1994)   Directions in conservation biology. *Journal of Animal Ecology,* **63**: 215–44.

Cavalieri, P. & Singer, P. (editors, 1993)   *The Great Ape Project.* Fourth Estate (London, UK).

CEC (1987)   *Mangroves of Africa and Madagascar: the Mangroves of Nigeria.* Prepared by the Societé d'Eco-Amenagement (SECA) and the Centre for Environmental Studies of the University of Leiden (CML). Commission of the European Communities (CEC, Brussels, Belgium).

CEC (1993)   *Environment Manual: (1) Environmental Procedures and Methodology Governing Lomé IV Development Co-operation Projects; (2) Sectoral Environmental Assessment Sourcebook.* Prepared by Environmental Resources Ltd for the Directorate-General for Development. Commission of the European Communities (CEC, Brussels, Belgium).

Chamber, R. (1983)   *Rural Development: Putting the Last First.* Longman (Harlow, UK).

Chambers, R. (1994a)   The origins and practice of participatory rural appraisal. *World Development.* **22**: 953–69.

Chambers, R. (1994b)   Participatory rural appraisal (PRA): analysis of experience. *World Development,* **22**: 1251–1268.

Chambers, R. (1994c)   Participatory rural appraisal (PRA): challenges, potentials and paradigm. *World Development,* **22**: 1437–1454.

Chin, S. C. (1985)   Agriculture and resource utilization in a lowland rainforest Kenyah community. *Sarawak Museum Journal, Special Monograph No. 3,* **35**: xvi+1–322.

Christy, F. T. (1982)   *Territorial Use Rights in Marine Fisheries: Definitions and Conditions.* FAO Fisheries Technical Paper 227. Food and Agriculture Organization of the United Nations (FAO, Rome, Italy).

Clark, J. (1992)   *Integrated management of coastal zones.* FAO Fisheries Technical Paper 327. Food and Agriculture Organization of the United Nations (FAO, Rome, Italy).

Clover, C. (1988)   Malaysia fury over western calls for timber boycott. *Daily Telegraph (London),* 19 April 1988.

CNPPA (1990)   *A Framework for the Classification of Terrestrial and Marine Protected Areas; Objectives, Criteria and Categories for Protected Areas.* Commission for National Parks and Protected Areas (CNPPA, IUCN, Gland, Switzerland).

Cochrane, J. (1992)   *Environmental Tourism Potential of Lore Lindu National Park and Morowali Nature Reserve.* The Nature Conservancy (TNC, Jakarta, Indonesia).

Coen, E. (1983)   Climate. Pp. 35–46 in *Costa Rican Natural History* (edited by D. H. Janzen). University of Chicago Press (Chicago, Illinois, USA).

Coghlan, A. (1994)   Costa Rican tree roots out plantation pests. *New Scientist,* 16 April 1994 (1921): 8.

Colchester, M. (1992) *Pirates, Squatters and Poachers: the Political Ecology of Dispossession of the Native Peoples of Sarawak.* Survival International (London, UK) and the Institute of Social Analysis (INSAN, Petaling Jaya, Malaysia).

Colfer, C. J. P., Chin, S. C., Peluso, N., Tamen Uyang Liq, Mackie, C., Matius, P., and Caldecott, J. O. (in press). Uma'Jalan Forestry. In *Beyond Slash and Burn: Lessons from the Kenyah on Improving Rainforest Management in Borneo* (edited by C. J. P. Colfer, S. C. Chin, and N. Peluso). Advances in Economic Botany Series. New York Botanical Garden (New York, USA).

Collins, M. (1980) *The Last Rain Forests: a World Conservation Atlas.* Oxford University Press (New York, USA).

Collins, N. M. & Morris, M. G. (1985) *Threatened Swallowtail Butterflies of the World: the IUCN Red Data Book.* World Conservation Union (IUCN, Gland, Switzerland).

Collins, N. M., Sayer, J. A., & Whitmore, T. C. (1991) *The Conservation Atlas of Tropical Forests – Asia and the Pacific.* Macmillan (London, UK).

Connor, J. & Martin, P. (1989) *Oban Project WWF Education Sector Study.* Appendix 8 of *Cross River National Park, Oban Division: Plan for Developing the Park and Its Support Zone.* World Wide Fund for Nature (WWF, Godalming, UK).

Cooke, A. (1994) *The Qoliqoli and Fiji – Management of Resources in Traditional Fishing Grounds.* Master's Dissertation, University of Newcastle-upon-Tyne (Newcastle, UK).

Corner, E. J. H. (1961) A tropical botanist's introduction to Borneo. *Sarawak Museum Journal,* **10**: 1–16.

Cox, C. R. (1988) *The Conservation Status of Biological Resources in the Philippines.* World Conservation Monitoring Centre (Cambridge, UK).

Cramb, R. A. & Wills, I. R. (1990) The role of traditional institutions in rural development: community-based land tenure and government land policy in Sarawak, Malaysia. *World Development,* **18**: 347–60.

Craven, I. (1989) The Arfak Mountains Nature Reserve, Bird's Head region, Irian Jaya, Indonesia. *Science in New Guinea,* **15**: 47–56.

Craven, I. & de Fretes, Y. (1987) *Arfak Mountains Nature Conservation Area: Management Plan, 1988–1992.* World Wide Fund for Nature (WWF, Jayapura, Irian Jaya, Indonesia).

Croft, J. R. (1992) *Summary Paper and Data Dictionary for Herbarium Information Standards and Protocols for Interchange of Data (HISPID).* Australian National Botanic Gardens (Canberra, Australia).

Cruz, R. V., Francisco, H. A, & Torres, C. S. (1991) *Agroecosystem Analysis of Makiling Forest Reserve, Philippines.* Report No. 1, Environment and Resources Management Project (College, Laguna, Philippines & Halifax, Nova Scotia, Canada).

Cultural Survival (1993) *State of the Peoples: a Global Human Rights Report on Societies in Danger.* Beacon Press (Boston, Massachusetts, USA).

Dahuri, R. (1994) Incorporating biodiversity objectives and criteria into environmental impact assessment laws and mechanisms in Indonesia. Pp. 319–25 in *Widening Perspectives in Biodiversity* (edited by A. F. Krattinger, J. A. McNeely, W. H. Kesser, K. R. Miller, Y. St Hill, & R. Senanayake). World Conservation Union (IUCN, Gland, Switzerland) & International Academy of the Environment (Geneva, Switzerland).

de Klemm, C. & Shine, C. (1993) *Biological Diversity Conservation and the Law: Legal Mechanisms for Conserving Species and Ecosystems.*

Environmental Policy and Law Paper No. 29. World Conservation Union (IUCN, Gland, Switzerland).

Diamond, J. (1986) The design of a nature reserve system for Indonesian New Guinea. Pp. 485–503 in *Conservation Biology: the Science of Scarcity and Diversity* (edited by M. E. Soulé). Sinauer (Sunderland, Massachusetts, USA).

Diamond, J. M. & May, R. M. (1976) Island biogeography and the design of natural reserves. Pp. 163–86 in *Theoretical Ecology: Principles and Applications* (edited by R. M. May). Blackwell (Oxford, UK).

Dick, J. (1991) *Forest Land Use, Forest Use Zonation, and Deforestation in Indonesia; a Summary and Interpretation of Existing Information.* Ministry of State for Population and Environment (KLH) & Environmental Impact Management Agency (BAPEDAL, Jakarta, Indonesia).

Dick, J. & Bailey, L. (1992) *Indonesia's Environmental Assessment Process (AMDAL): Progress, Problems and a Suggested Blueprint for Improvement.* Environmental Management Development in Indonesia Project (EMDI, Jakarta, Indonesia & Halifax, Nova Scotia, Canada).

Dixon, J. A., Scura, L. F., & van't Hof, T. (1993) Meeting ecological and economic goals: marine parks in the Caribbean. *Ambio,* **22**: 117–25.

Dizon, J. T. (1992) *Socio-Economic Profile of Farmers in the Makiling Forest Reserve.* College of Forestry & Environment and Resource Management Project, University of the Philippines at Los Baños (UPLB College, the Philippines).

Dompka, V. (1994) *Population and the Environment, a WWF Discussion Document.* World Wide Fund for Nature (WWF, Gland, Switzerland).

Dourojeanni, M. J. (1985) Over-exploited and under-used animals in the Amazon region. Pp. 419–33 in *Key Environments: Amazonia* (edited by G. T. Prance & T. E. Lovejoy). Pergamon (Oxford, UK).

Dransfield, J. (1979) *A Manual of the Rattans of the Malay Peninsula.* Forest Department, Ministry of Primary Industries Malaysia (Kuala Lumpur, Malaysia).

Dransfield, J. (1984) *The Rattans of Sabah.* Sabah Forest Record No. 13. Forest Department (Sandakan, Sabah, Malaysia).

DtE (1991) *Pulping the Rainforest: the Rise of Indonesia's Paper and Pulp Industries.* Down to Earth, Asia-Pacific People's Environment Network (Penang, Malaysia).

Eber, S. (editor, 1992) *Beyond the Green Horizon: Principles for Sustainable Tourism.* Tourism Concern (London, UK) and World Wide Fund for Nature (WWF, Godalming, UK).

EC (1988) Resolution on the catastrophic environmental impact of large-scale deforestation in Sarawak (East Malaysia). *Official Journal of the European Communities,* 31(C235):196–8.

EC (1990) Dossier: Tourism. *The Courier (Africa-Caribbean-Pacific & European Community),* **122**: 50–86.

Eisner, T. (1990) Prospecting for Nature's Chemical Riches. *Issues in Science and Technology,* Winter 1989–90: 331–4.

Erokoro, P., White, J. & Caldecott, J. O. (1990) *Management Options for the Obudu Plateau.* Appendix 10 of *Cross River National Park, Okwangwo Division: Plan for Developing the Park and Its Support Zone.* World Wide Fund for Nature (WWF, Godalming, UK).

FAO (1993) *FAO Forest Products Yearbook, 1980–1991.* FAO Forestry Series No. 26/FAO Statistics Series No. 110. Food and Agriculture Organization of the United Nations (Rome, Italy).

Farnsworth, N. R. (1988)   Screening plants for new medicines. Pp. 83–97 in
    Biodiversity (edited by E. O. Wilson). National Academy Press (Washington
    DC, USA).
FGN (1989)   *Re: Dedicated Sovereign Debt Servicing for the Management of
    Cross River National Park and Its Support Zone.* Letter F-12310/S.5/23 of the
    28 August 1989 to the Nigerian Conservation Foundation (NCF) and the
    World Wide Fund for Nature (WWF). Federal Ministry of Finance and
    Economic Development (Lagos, Nigeria).
FGN (1991) Decree No. 36: National Parks Decree 1991.   *Federal Republic of
    Nigeria Official Gazette (Lagos, 26th August 1991),* 78(44): A213–63.
Filion, F. L., Foley, J. P., & Jacquemot, A. J. (1994)   The economics of global
    ecotourism. Pp. 235–52 in *Protected Area Economics and Policy: Linking
    Conservation and Sustainable Development* (edited by M. Munasinghe & J.
    McNeely). The World Bank (Washington, DC, USA) & World Conservation
    Union (IUCN, Gland, Switzerland).
Fisher, R. & Ury, W. (1981)   *Getting to Yes: Negotiating Agreement Without
    Giving In.* Hutchinson (London, UK).
Forrest, T. (1993)   *Politics and Economic Development in Nigeria.* Westview
    (Boulder, Colorado, USA).
Fox, J. M. (1989)   Diagnostic tools for social forestry. *Journal of World Forest
    Resources Management,* 4: 61–77.
Francisco, H. A., Mallion, F. K., & Sumalde, Z. M. (1992)   *Economics of
    Dominant Forest Based Cropping Systems in Mt Makiling Forest Reserve: an
    Integrative Report.* Institute of Environmental Science and Management,
    University of the Philippines at Los Baños (UPLB, College, the Philippines).
Frankel, O. H. & Soulé, M. E. (1981)   *Conservation and Evolution.* Cambridge
    University Press (Cambridge, UK).
Freeman, D. (1970)   *Report on the Iban.* London School of Economics
    Monographs on Social Anthropology No. 41. Athlone (London, UK).
Freire, P. (1970)   *Pedagogy of the Oppressed.* Translated by M. B. Ramos.
    Continuum (New York, USA).
FUAM (1994)   *Blue Dolphin of Malta Underwater Photographic Competition*
    (Press release, 24 October 1994). Federation of Underwater Activities of
    Malta (Valetta, Malta).
Gadsby, E. L. (1990)   *The Status and Distribution of the Drill (*Mandrillus
    leucophaeus*) in Nigeria.* Wildlife Conservation Society (WCS, New York,
    USA) & World Wide Fund for Nature (WWF, Washington, DC, USA).
Gadsby, E. L. & Jenkins, P. D., Jr (1995)   The FFPS drill project. *Fauna and
    Flora News,* April 1995: 1–2.
Gage, J. D. & Tyler, P. A. (1991)   *Deep-Sea Biology: a Natural History of
    Organisms at the Deep-Sea Floor.* Cambridge University Press (Cambridge,
    UK).
Gámez, R. (1991a)   Development, preservation of tropical biological diversity, and
    the case of Costa Rica. *International Agriculture Newsletter,* January 1991: 1–
    3.
Gámez, R. (1991b)   Biodiversity conservation through facilitation of its
    sustainable use: Costa Rica's National Biodiversity Institute. *Trends in
    Ecology and Evolution,* 6: 377–8.
Gámez, R., Piva, A., Sittenfeld, A., Leon, E., Jiminez, J. & Mirabelli, G. (1993)
    Costa Rica's conservation program and National Biodiversity Institute
    (INBio). Pp. 53–67 in *Biodiversity Prospecting* (edited by W. V. Reid, A.

Sittenfeld, S. A. Laird, D. H. Janzen, C. A. Meyer, M. A. Gollin, R. Gámez, & C. Juma). World Resources Institute (Washington, DC, USA).

Garcia, S. M. (1994)   The precautionary principle: its implications in capture fisheries management. *Ocean & Coastal Management*, **22**: 99–125.

GEF (1994)   *Quarterly Operational Report, November 1994*. Global Environment Facility Secretariat (Washington, DC, USA).

Ghazi, P., Smith, F. & Trevena, C. (1995)   The rape of the oceans. The *Observer* (London), 2 April 1995: 23.

Gillis, M. (1988a)   Malaysia: public policies and the tropical forest. Pp. 115–64 in *Public Policies and the Misuse of Forest Resources* (edited by R. Repetto & M. Gillis). Cambridge University Press (Cambridge, UK).

Gillis, M. (1988b)   Indonesia: public policies, resource management, and the tropical forest. Pp. 43–113 in *Public Policies and the Misuse of Forest Resources* (edited by R. Repetto & M. Gillis). Cambridge University Press (Cambridge, UK).

Glowka, L., Burhenne-Guilmin, F., Synge, H., McNeely, J. A., & Gündling, L. (1994)   *A Guide to the Convention on Biological Diversity*. Environmental Policy and Law Paper No. 30. World Conservation Union (IUCN, Gland, Switzerland).

Gollin, M. A. (1993)   An intellectual property rights framework for biodiversity prospecting. Pp. 159–97 in *Biodiversity Prospecting* (edited by W. V. Reid, A. Sittenfeld, S. A. Laird, D. H. Janzen, C. A. Meyer, M. A. Gollin, R. Gámez, & C. Juma). World Resources Institute (Washington, DC, USA).

Gressitt, J. L. (editor, 1982)   *Ecology and Biogeography of New Guinea*. W. Junk (The Hague, the Netherlands).

Haddad, W. D., Carnoy, M., Rinaldi, R., & Regel, O. (1990)   *Education and Development: Evidence for New Priorities*. World Bank Discussion Papers, 95. The World Bank (Washington, DC, USA).

Hamilton, L. S. & Snedaker, S. C. (1984)   *Handbook for Mangrove Area Management*. Environment and Policy Institute, East–West Centre (Honolulu, Hawai'i, USA).

Hardjono, J. (1991)   The dimensions of Indonesia's environmental problems. Pp. 1–16 in *Indonesia – Resources, Ecology, and Environment* (edited by J. Hardjono). Oxford University Press (New York, USA).

Harrisson, T. (1956)   Rhinoceros in Borneo: and traded to China. *Sarawak Museum Journal*, **7**: 263–74.

Hartshorn, G. S. (1983a)   Plants: introduction. Pp. 118–57 in *Costa Rican Natural History* (edited by D. H. Janzen). University of Chicago Press (Chicago, Illinois, USA).

Hartshorn, G. S. (1983b)   Wildlands conservation in Central America. Pp. 423–44 in: *Tropical Rain Forest: Ecology and Management* (edited by S. L. Sutton, T. C. Whitmore, & A. C. Chadwick). Blackwell (Oxford, UK).

Hatch, T. (1982)   *Shifting Cultivation in Sarawak – a Review*. Soils Division, Research Branch, Department of Agriculture (Kuching, Sarawak, Malaysia).

Hay, J. E. (editor, 1992)   *Ecotourism Business in the Pacific: Promoting a Sustainable Experience*. University of Aukland (Aukland, New Zealand) & East–West Center (Honolulu, Hawai'i, USA).

Heaney, L. R. (1986)   Biogeography of mammals in SE Asia: estimates of rates of colonization, extinction and speciation. *Biological Journal of the Linnean Society*, **28**: 127–65.

Herz, B., Subbarao, K., Habib, M., & Raney, L. (1991)   *Letting Girls Learn: Promising Approaches in Primary and Secondary Education*. World Bank Discussion Papers, 133. The World Bank (Washington, DC, USA).

Hislop, J. A. (1949) Some field notes on the bearded pig. *Malayan Nature Journal*, 4: 62–4.

Hislop, J. A. (1952) More about the bearded pig. *Malayan Nature Journal*, 7: 22–3.

Hislop, J. A. (1955) Notes on the migration of the bearded pig. In *Excavations at Gua Cha, Kelantan, 1954, Appendix E* (edited by G. de G. Sieveking). *Federation Museums Journal (*New Series), 1–2: 134–7.

Holdgate, M. W. (1993) *Concept Paper: Conservation of Biological Diversity in Hainan Province.* World Conservation Union (IUCN, Gland, Switzerland).

Holland, M. D., Allen, R. K. G., Barton, D., & Murphy, S. T. (1989) *Cross River National Park, Oban Division: Land Evaluation and Agricultural Recommendations.* Natural Resources Institute (NRI, Chatham, UK).

Hollis, G. E., Adams, W. M., & Aminu Kano, M. (editors, 1993) *The Hadejia-Nguru Wetlands: the Environment, Economy and Sustainable Development of a Sahelian Floodplain Wetland.* World Conservation Union (IUCN, Gland, Switzerland).

Hunt, L. (1992) *Integrated Resource Plan, Kabupaten Manokwari, Province of Irian Jaya.* Integrated Regional Environmental Development Programme (INREDEP), Ministry of State for Population and Environment (KLH, Jakarta, Indonesia).

Hurst, F. & Thompson, H. (1994) *A Review of the Okwangwo Division of the Cross River National Park.* Final Report, 26 July 1994. World Wide Fund for Nature (WWF, Godalming, UK).

Huston, M. A. (1994) *Biological Diversity: the Coexistence of Species on Changing Landscapes.* Cambridge University Press (Cambridge, UK).

ICBP (1992) *Putting Biodiversity on the Map: Priority Areas for Global Conservation.* Birdlife International (ICBP, Cambridge, UK).

ICBP (1995) *The ICBP Directory of Endemic Bird Areas.* BirdLife International (Cambridge, UK).

IISD (1994a) *Trade and Sustainable Development Principles.* International Institute for Sustainable Development (IISD, Winnipeg, Manitoba, Canada).

IISD (1994b) *GATT, the WTO and Sustainable Development: Positioning the Work Program on Trade and Environment.* International Institute for Sustainable Development (IISD, Winnipeg, Manitoba, Canada).

IPCC (1990) *Climate Change: the IPCC Scientific Assessment* (edited by J. T. Houghton, G. J. Jenkins, & J. J. Ephraums for the Intergovernmental Panel on Climate Change). Cambridge University Press (Cambridge, UK).

IUCN (1987a) *Action Strategy for Protected Areas in the Afrotropical Realm.* World Conservation Union (IUCN, Gland, Switzerland).

IUCN (1987b) *Directory of Afrotropical Protected Areas.* World Conservation Union (IUCN, Gland, Switzerland).

IUCN (1992) *Improving Management in and Around Protected Areas: an Investment Framework.* Draft prepared by the Commission on National Parks & Protected Areas (CNPPA), World Conservation Monitoring Centre (WCMC), IUCN-US and the Centre for Social & Economic Research on the Global Environment (CSERGE). World Conservation Union (IUCN, Gland, Switzerland).

IUCN (1993) *Draft Guidelines for the Ecological Sustainability of Non-Consumptive and Consumptive Uses of Wild Species.* World Conservation Union (IUCN, Gland, Switzerland).

IUCN (1994) *Draft Initial Procedure for Assessing the Sustainability of Use of Wild Species.* World Conservation Union (IUCN, Gland, Switzerland).

IUCN, UNEP, & WWF (1980) *World Conservation Strategy.* World Conservation Union (IUCN), United Nations Environment Programme (UNEP) and World Wide Fund for Nature (WWF). World Conservation Union (IUCN, Gland, Switzerland).

IUCN, UNEP, & WWF (1991) *Caring for the Earth: a Strategy for Sustainable Living.* World Conservation Union (IUCN), United Nations Environment Programme (UNEP) and World Wide Fund for Nature (WWF). World Conservation Union (IUCN, Gland, Switzerland).

Ité, U. E. (1995) *Agriculture and Tropical Forest Conservation in South East Nigeria.* Doctoral dissertation, University of Cambridge (Cambridge, UK).

ITTO (1990) *The Promotion of Sustainable Forest Management: a Case Study in Sarawak, Malaysia.* International Tropical Timber Organization (ITTO, Yokohama, Japan).

Janzen, D. H. (1974) Tropical blackwater rivers, animals, and mast fruiting by the Dipterocarpaceae. *Biotropica,* **6**: 69–103.

Janzen, D. H. (1975) *Ecology of Plants in the Tropics.* Arnold (London, UK).

Janzen, D. H. (editor, 1983) *Costa Rican Natural History.* University of Chicago Press (Chicago, Illinois, USA).

Janzen, D. H. (1986a) The future of tropical ecology. *Annual Review of Ecology and Systematics,* **17**: 305–324.

Janzen, D. H. (1986b) The eternal external threat. Pp. 286–303 in *Conservation Biology: the Science of Scarcity and Diversity* (edited by M. E. Soulé). Sinauer (Sunderland, Massachusetts, USA).

Janzen, D. H. (1988a) Tropical dry forests: the most endangered major tropical ecosystems. Pp. 130–7 in *Biodiversity* (edited by E. O. Wilson). National Academy Press (Washington, DC, USA).

Janzen, D. H. (1988b) Guanacaste National Park: tropical ecological and biocultural restoration. Pp. 143–92 in *Rehabilitating Damaged Ecosystems* (edited by J. Cairns Jr). CRC Press (Boca Raton, Florida, USA).

Janzen, D. H. (1991) The National Biodiversity Institute of Costa Rica: how to save tropical biodiversity. *American Entomologist.* Fall 1991: 159–71.

Janzen, D. H. (1992a) A south-north perspective on science in the management, use, and economic development of biodiversity. Pp. 27–52 in *Conservation of Biodiversity for Sustainable Development* (edited by O. T. Sandlund, K. Hindar, & A. H. D. Brown). Scandinavian University Press (Universitetsforlaget, Oslo, Norway).

Janzen, D. H. (1992b) *Inventory 1999: A Process in the Early Stages of Its Evolution.* Draft proposal to the United States National Science Foundation (13 November 1992).

Janzen, D. H. (1994) Priorities in tropical biology. *Trends in Ecology and Evolution,* **9**: 365–8.

Janzen, D. H., Hallwachs, W., Jiminez, J., & Gámez, R. (1993) The role of parataxonomists, inventory managers, and taxonomists in Costa Rica's national biodiversity inventory. Pp. 223–54 in *Biodiversity Prospecting* (edited by W. V. Reid, A. Sittenfeld, S. A. Laird, D. H. Janzen, C. A. Meyer, M. A. Gollin, R. Gámez, & C. Juma). World Resources Institute (Washington, DC, USA).

Janzen, D. H., Hallwachs, W., Gámez, R., Sittenfeld, A. & Jiminez, J. (1993) Research management policies: permits for collecting and research in the tropics. Pp. 131–57 in *Biodiversity Prospecting* (edited by W. V. Reid, A. Sittenfeld, S. A. Laird, D. H. Janzen, C. A. Meyer, M. A. Gollin, R. Gámez, & C. Juma). World Resources Institute (Washington, DC, USA).

Jermy, A. C. & Kavanagh, K. P. (editors, 1982)   Gunung Mulu National Park, Sarawak. *Sarawak Museum Journal*, **30**, Special Issue 2, Part I: 1–279.

Jermy, A. C. & Kavanagh, K. P. (editors, 1984)   Gunung Mulu National Park, Sarawak. *Sarawak Museum Journal*, **30**, Special Issue 2, Part II:1–190.

Johannes, R. E. (1981)   *Words of the Lagoon: Fishing and Marine Lore in the Palau District of Micronesia.* University of California Press (Berkeley, California, USA).

Johannes, R. E. & Hatcher, B. G. (1986)   Shallow tropical marine environments. Pp. 371–93 in *Conservation Biology: the Science of Scarcity and Diversity* (edited by M. E. Soulé). Sinaur (Sunderland, Massachusetts, USA).

John Paul II, His Holiness the Pope. (1989)   *Message for the World Day of Peace 1990.* The Vatican (Rome, Italy).

John Paul II, His Holiness the Pope. (1990)   *Address Concerning Tropical Forests and the Conservation of Species.* Pontifical Academy of Sciences, The Vatican (Rome, Italy).

Jong, K., Stone, B. C., & Soepadmo, E. (1973)   Malaysian tropical forest: an underexploited genetic reservoir of edible-fruit tree species. Pp. 113–21 in *Proceedings of the Symposium on Biological Resources & National Development* (edited by E. Soepadmo & K. G. Singh). Malayan Nature Society (Kuala Lumpur, Malaysia).

Joyce, C. (1991)   Prospectors for tropical medicines. *New Scientist*, 19 October 1991: 36–40.

Kang, B. T., Wilson, G. F., & Lawson, T. L. (1984)   *Alley Cropping, a Stable Alternative to Shifting Cultivation.* International Institute for Tropical Agriculture (IITA, Ibadan, Nigeria).

Kang, B. T., Van der Kruis, A. C. B. M., & Couper, D. C. (1986)   Alley cropping for food crop production in the humid and subhumid tropics. Pp. 16–26 in *Alley Farming in the Humid and Subhumid Tropics* (edited by B. T. Kang & L. Reynolds). International Institute for Tropical Agriculture (IITA, Ibadan, Nigeria).

Kavanagh, M. & Caldecott, J. O. (1988)   Guidelines for the translocation of wild primates. In *Active Management for the Conservation of Wild Primates* (edited by W. Konstant and A. Mitchell). IUCN-SSC Second Occasional Paper. World Conservation Union (IUCN, Gland, Switzerland).

Kavanagh, M., Abdul Rahim, A., & Hails, C. J. (1989)   *Rainforest Conservation in Sarawak: an International Policy for WWF.* World Wide Fund for Nature (WWF, Kuala Lumpur, Malaysia and Gland, Switzerland).

Kelleher, G. & Bleakley, C. (1992)   *A Global Representative System of Marine Protected Areas: Report to the World Bank Environment Department.* Great Barrier Reef Marine Park Authority (Canberra, Australia) and World Conservation Union (IUCN, Gland, Switzerland).

Kemf, E. (editor, 1993)   *The Law of the Mother: Protecting Indigenous Peoples in Protected Areas.* Sierra Club (San Francisco, California, USA).

Kempe, J. E. (1948)   The riddle of the bearded pig. *Malayan Nature Journal*, **3**: 36–42.

Kimmage, K. & Adams, W. M. (1992)   Wetland agricultural production and river basin development in the Hadejia-Jama'are valley, Nigeria. *The Geographical Journal*, **158**: 1–12.

King, V. T. (1978)   Introduction. Pp. 1–36 in *Essays on Borneo Societies* (edited by V. T. King). Oxford University Press (Oxford, UK).

Kiss, A. (editor, 1990)   *Living with Wildlife: Wildlife Resource Management with Local Participation in Africa.* The World Bank (Washington, DC, USA).

KLH (1992) *Indonesian Country Study on Biological Diversity* (edited by the Indonesian Biodiversity Country Study Standing Committee). Also published in Indonesian. Ministry of State for Population and Environment (KLH, Jakarta, Indonesia).

Knight, W. (1989) *Education Sector Review*. Appendix 9 of *Cross River National Park, Oban Division: Plan for Developing the Park and Its Support Zone*. World Wide Fund for Nature (WWF, Godalming, UK).

Kottelat, M., Whitten, A. J., Kartikasari, S. N., & Wirjoatmodjo, S. (1993) *Fresh-water Fishes of Western Indonesia and Sulawesi*. Also published in Indonesian. Periplus (Singapore).

Kristiansen, M. S., Lamoureux, C. H., Stone, B. C., Woolliams, K. R., & Madulid, D. A. (1993) Ex-situ *Conservation of the Plant Resources of the Philippines: an Emergency Rescue Operation to Develop Living Collections of Threatened Philippine Plants*. Honolulu Botanical Gardens (Honolulu, Hawai'i, USA)

Krumpe, E. E. & McCoy, L. (1992) Techniques to resolve conflict in natural resource management in parks and protected areas. In *Papers on Conflict Resolution From the IVth World Congress on National Parks and Protected Areas*. World Conservation Union (IUCN, Gland, Switzerland) & Keystone Centre (Keystone, Colorado, USA).

Kuiter, R. (1992) *Tropical Reef-Fishes of the Western Pacific: Indonesia and Adjacent Waters*. Gramedia Pustaka Utama (Jakarta, Indonesia).

Laffoley, D. (1995) Tides of change? *The Biologist*, 42(1): 13–16.

Laidlaw, R. K. (1994) *The Virgin Jungle Reserves of Peninsular Malaysia: the Ecology and Dynamics of Small Protected Areas in Managed Forest*. Doctoral Dissertation, University of Cambridge (Cambridge, UK).

Laird, S.A (1993) Contracts for biodiversity prospecting. Pp. 99–130 in *Biodiversity Prospecting* (edited by W. V. Reid, A. Sittenfeld, S. A. Laird, D. H. Janzen, C. A. Meyer, M. A. Gollin, R. Gámez, & C. Juma). World Resources Institute (Washington, DC, USA).

Lapidaire, J. M., Buchmann, K. E., Saeger, J., Davies, G., & Thill, M. (1991) *Samar Province Agricultural Resources Development Programme*. SETA and Commission of the European Communities (Brussels, Belgium)

Laurance, W. F. (1991) Ecological correlates of extinction proneness in Australian tropical rain forest mammals. *Conservation Biology*, 5(1): 79–89.

Leaman, D. J., Yusuf, R., & Sangat-Roemantyo, H. (1991) *Kenyah Dayak Forest Medicines*. World Wide Fund for Nature (WWF, Jakarta, Indonesia).

Lean, G. (1995) The empty oceans. *The Independent on Sunday (London)*, 15 January 1995: 17.

Lee, P. C., Thornback, J. & Bennett, E. L. (1988) *Threatened Primates of Africa: the IUCN Red Data Book*. World Conservation Union (IUCN, Gland, Switzerland).

Lewington, A. (1991) Eat more Brazil nuts! *WWF Tropical Forests Update*. January 1991: 4.

Lewis, C. (1992) Parks and people in conflict: a framework for analysis and action. In *Papers on Conflict Resolution From the IVth World Congress on National Parks and Protected Areas*. World Conservation Union (IUCN, Gland, Switzerland) & Keystone Centre (Keystone, Colorado, USA).

Lewis, C. (1993) Nature in the crossfire. Pp. 123–30 in *The Law of the Mother: Protecting Indigenous Peoples in Protected Areas* (edited by E. Kemf). Sierra Club (San Francisco, California, USA).

Lieske, E. & Myers, R. (1994) *Coral Reef Fishes: Indo-Pacific & Caribbean*. HarperCollins (London, UK).

Li, J-C., Kong, F-W., He, N-H., & Ross, L. (1988) Price and policy: the keys to revamping China's forestry resources. Pp. 205–45 in *Public Policies and the Misuse of Forest Resources* (edited by R. Repetto & M. Gillis). Cambridge University Press (Cambridge, UK).

Li, T. M. (1993) *Gender Issues in Community-Based Resource Management: Theories, Applications and Philippine Case Studies.* Report No. 9, Environment and Resources Management Project (College, Laguna, Philippines & Halifax, Nova Scotia, Canada).

Liu, D-H. (1992) *Protecting Mt Jianfeng Ling and Building the Earth Village – Progress and Prospect on the Subject on the Conservation and Development of the Tropical Rain Forests in Mt Jianfeng Ling.* International Advisory Council on the Economic Development of Hainan in Harmony with the Natural Environment (Haikou, Hainan, China).

Lovejoy, T. E., Bierregaard, R. O. Jr., Rylands, A. B., Malcolm, J. R., Quintela, C. E., Harper, L. H., Brown, K. S. Jr., Powell, A. H., Powell, G. V. N., Shubart, H. O. R., & Hays, M. B. (1986) Edge and other effects of isolation on Amazon forest fragments. Pp. 257–85 in *Conservation Biology: the Science of Scarcity and Diversity* (edited by M. E. Soulé). Sinaur (Sunderland, Massachusetts, USA).

Lowe, R. G. (1986) *Agricultural Revolution in Africa?* Macmillan (London, UK).

Lucas, P. H. C. (1990) *Protected Landscapes: a Guide for Policy Makers and Planners.* World Conservation Union (IUCN, Gland, Switzerland).

Lyons, S. (1991) Research pact may help rain forests pay for their keep: Merck, Costa Rica team up to prospect for new drug chemicals. *The Boston Globe*, 4 November 1991.

MacDonald, A. B. (1992) *File Note Re: Philippines, Local Government Code.* Delegation of the Commission of the European Communities (Manila, Philippines).

MacKinnon, J. R. (1994a) *Analytical Status Report of Biodiversity Conservation in the Asia-Pacific Region.* Asian Bureau for Conservation (Hong Kong, China).

MacKinnon, J. R. (1994b) *Too Many Plans: Personal Views on the Planning Process.* Asian Bureau for Conservation (Hong Kong, China) and World Conservation Union (IUCN, Gland, Switzerland).

MacKinnon, J. R. & Artha, B. (1981–2) *National Conservation Plan for Indonesia* (8 volumes: 1 – Introduction; 2 – Sumatra; 3 – Java and Bali; 4 – Lesser Sundas; 5 – Kalimantan; 6 – Sulawesi; 7 – Maluku and Irian Jaya; 8 – General Topics). Food and Agriculture Organization of the United Nations (FAO, Bogor, Indonesia).

MacKinnon, J. R. & MacKinnon, K. S. (1986a) *Review of the Protected Areas System in the Afrotropical Realm.* World Conservation Union (IUCN, Gland, Switzerland) & United Nations Environment Programme (UNEP, Nairobi, Kenya).

MacKinnon, J. R. & MacKinnon, K. S. (1986b) *Review of the Protected Area System in the Indo-Malayan Realm.* World Conservation Union (IUCN, Gland, Switzerland) & United Nations Environment Programme (UNEP, Nairobi, Kenya).

MacKinnon, J. R. & MacKinnon, K. S. (1986c) *Review of the Protected Area System in Oceania.* World Conservation Union (IUCN, Gland, Switzerland) & United Nations Environment Programme (UNEP, Nairobi, Kenya).

MacKinnon, J. R., MacKinnon, K. S., Child, G., & Thorsell, J. (1986) *Managing Protected Areas in the Tropics.* World Conservation Union (IUCN, Gland, Switzerland).

MacKinnon, K. S., Hatta, G., Halim, H., & Mangalik, A. (in press) *The Ecology of Kalimantan.* Periplus (Singapore).

Madulid, D. A. (1982) Plants in peril. *The Philippines Journal of Science and Culture,* **3**: 8–16.

Magrath, W. B. & Arens, P. (1987) *The Costs of Soil Erosion on Java – A Natural Resource Accounting Approach.* World Resources Institute (WRI, Washington, DC, USA).

Mallion, F. K., Francisco, H. A., & Gagalac, M. A. (1992) *Evolution of Dominant Forest Based Cropping Systems in the Mt Makiling Forest Reserve.* Environment and Resources Management Project (College, Laguna, Philippines & Halifax, Nova Scotia, Canada).

Manembu, N. (1991) *The Sempan, Nduga, Nakai, and Amungme Peoples of the Lorentz Area.* World Wide Fund for Nature (Jakarta, Indonesia).

Mangel, M. & Tier, C. (1994) Four facts every conservation biologist should know about persistence. *Ecology,* **75**: 607–14.

Marsh, C. W. & Wilson, W. L. (1981) *A Survey of Primates in Peninsular Malaysian Forests.* Malaysian Primates Research Programme, National University of Malaysia (UKM, Kuala Lumpur, Malaysia).

Marshall, P. (1993a) *Cross River National Park, Oban Hills Integrated Rainforest Conservation Programme NGO001 (formerly 3264): Report for the Period 1st April 1992 to 31st March 1993.* World Wide Fund for Nature (WWF, Godalming, UK).

Marshall, P. (1993b) *Cross River National Park, Okwangwo Division, Development of the Northern Sector of Cross River National Park NGO003 (formerly 3916): Report for the Period 1st April 1992 to 31st March 1993.* World Wide Fund for Nature (WWF, Godalming, UK).

MASKAYU (1994) *Monthly Timber Bulletin of the Malaysian Timber Industry Board.* Ministry of Primary Industries (Kuala Lumpur, Malaysia).

MBEF (1994) *Strategy for the Rehabilitation and Management of the Maqueda Bay Ecosystem.* Maqueda Bay Ecosystem Forum (Catbalogan, Samar, Philippines).

McCarthy, J. (1991) *Conservation Areas of Indonesia.* World Conservation Monitoring Centre (Cambridge, UK).

McCracken, J. A., Pretty, J. N., & Conway, G. R. (1988) *An Introduction to Rapid Rural Appraisal for Agricultural Development.* International Institute for Environment and Development (IIED, London, UK).

McKie, R. (1994) Parrot-fish and chips, please. *The Observer (London),* 7 August 1994: 3.

McNeely, J. A. (1988) *Economics and Biological Diversity.* World Conservation Union (IUCN, Gland, Switzerland).

Miles, S., Teleki, G. C., Crawford, A., & Pleszczynska, K. (1993) *EMDI Phase III – Environmental Management Development in Indonesia: Mid-Term Evaluation Report.* Canadian International Development Agency (Ottawa, Canada).

Mitchell, A. H. (1982) *Siberut Nature Conservation Area Management Plan, 1983–1988.* World Wide Fund for Nature Indonesia Program (Bogor, Indonesia).

MoF & FAO (1991) *Indonesian Tropical Forestry Action Programme.* Ministry of Forestry and the Food and Agriculture Organization of the United Nations (Jakarta, Indonesia).

Monk, K., de Fretes, Y., & Lilley G. (in press) *The Ecology of Nusa Tenggara and Maluku.* Periplus (Singapore).

Morakinyo, A. B. (1994) *The Ecology and Silviculture of Rattans in Africa: a Management Strategy for Cross River State and Edo State, Nigeria.* Master's Dissertation, University College of North Wales (Bangor, UK).

Muller, K. (1990) *Indonesian New Guinea: Irian Jaya.* Periplus (Singapore).

Muller, K. (1992) *Underwater Indonesia: a Guide to the World's Greatest Diving.* Periplus (Singapore).

Munasinghe, M. (1993) *Environmental Economics and Sustainable Development.* World Bank Environment Paper No. 3. The World Bank (Washington, DC, USA).

Munasinghe, M. (1994) Economic and policy issues in natural habitats and protected areas. Pp. 15–49 in *Protected Area Economics and Policy: Linking Conservation and Sustainable Development* (edited by M. Munasinghe & J. McNeely). The World Bank (Washington, DC, USA) and World Conservation Union (IUCN, Gland, Switzerland).

Munasinghe, M. & McNeely, J. (editors, 1994) *Protected Area Economics and Policy: Linking Conservation and Sustainable Development.* The World Bank (Washington, DC, USA) and World Conservation Union (IUCN, Gland, Switzerland).

Mundy, J. R. J. H. (1989) *Nigeria, Oban Hills National Park: Analysis of Funding Options.* World Wide Fund for Nature (WWF, Godalming, UK).

Myers, N. (1984) *The Primary Source.* Norton (New York, USA).

Myers, N. (1986) Tropical deforestation and a mega-extinction spasm. Pp. 394–426 in *Conservation Biology: the Science of Scarcity and Diversity* (edited by M. E. Soulé). Sinaur (Sunderland, Massachusetts, USA).

Myers, N. (1988) Tropical forests: much more than stocks of wood. *Journal of Tropical Ecology*, **4**: 209–21.

Myers, R. F. (1991) *Micronesian Reef Fishes: a Practical Guide to the Identification of the Coral Reef Fishes of the Tropical Central and Western Pacific.* Coral Graphics (Guam, USA).

Nash, S. V. (1991) *The WWF Irian Jaya Conservation Programme: 3-Year Evaluation Summary and Proposed 1991–1992 Work Plan.* World Wide Fund for Nature (WWF, Jayapura, Indonesia).

Nectoux, F. & Kuroda, Y. (1989) *Timber from the South Seas: an Analysis of Japan's Tropical Timber Trade and its Environmental Impact.* World Wide Fund for Nature (WWF, Gland, Switzerland).

Neville, D. (1992) *Butterfly Farming in the Arfak Mountains of Irian Jaya, a Project Update.* World Wide Fund for Nature (Manokwari, Irian Jaya, Indonesia).

New, T. R. (1994) Butterfly ranching: a sustainable use of insects and sustainable benefit to habitats. *Oryx*, **28**: 169–72.

Nielsen, L. & Brown, R. D. (editors, 1988) *Translocation of Wild Animals.* Wisconsin Humane Society (Milwaukee, Wisconsin, USA) and Caesar Kleberg Wildlife Research Institute (Kingsville, Texas, USA).

Nightingale, N. (1992) *New Guinea: an Island Apart.* BBC Books (London, UK).

Nolledo, J. N. (1993) *The Local Government Code of 1991, Annotated 1993 Reprint with August, 1993 Addendum.* National Book Store (Manila, Philippines).

Norse, E. A. (editor, 1993) *Global Marine Biological Diversity: a Strategy for Building Conservation into Decision Making.* Island Press (Washington, DC, USA).

NPWO (1984) *A Proposal to Constitute Pulong Tau National Park in the Fourth and Fifth Divisions of Sarawak.* National Parks & Wildlife Office, Sarawak Forest Department (Kuching, Sarawak, Malaysia).

NPWO (1987a)   *A Revised Proposal to Constitute the Pulong Tau National Park in the Miri and Limbang Divisions of Sarawak.* National Parks & Wildlife Office, Sarawak Forest Department (Kuching, Sarawak, Malaysia).

NPWO (1987b)   *A Proposal to Constitute the Usun Apau National Park in the Miri and Kapit Divisions of Sarawak.* National Parks & Wildlife Office, Sarawak Forest Department (Kuching, Sarawak, Malaysia).

NPWO (1987c)   *A Proposal to Constitute the Dulit Range National Park in the Miri and Kapit Divisions of Sarawak.* National Parks & Wildlife Office, Sarawak Forest Department (Kuching, Sarawak, Malaysia).

NRC (1993)   *Vetiver Grass: a Thin Green Line Against Erosion.* National Research Council, National Academy Press (Washington, DC, USA).

Oates, J. F. (1995)   The dangers of conservation by rural development – a case-study from the forests of Nigeria. *Oryx,* **29**: 115–22.

Oates, J. F., White, D., Gadsby, E. L., & Bisong, P. (1990)   *Conservation of Gorillas and Other Species.* Appendix 1 of *Cross River National Park, Okwangwo Division: Plan for Developing the Park and Its Support Zone.* World Wide Fund for Nature (WWF, Godalming, UK).

ODNRI (1989)   *Nigeria: Profile of Agricultural Potential.* Overseas Development Natural Resources Institute (ODNRI, Chatham, UK).

Okafor, J. C. (1978)   Development of forest tree crops for food supplies in Nigeria. *Forest Ecology & Management,* **1**: 235–47.

Okafor, J. C. (1979)   Edible indigenous woody plants in the rural economy of the Nigerian forest zone. Pp. 262–300 in *The Nigerian Rainforest Ecosystem* (edited by D. U. U. Okali). University of Ibadan Press (Ibadan, Nigeria).

Okafor, J. C. (1989)   *Agroforestry Aspects.* Appendix 2 of *Cross River National Park, Oban Division: Plan for Developing the Park and Its Support Zone.* World Wide Fund for Nature (WWF, Godalming, UK).

Okafor, J. C. (1990)   *Agroforestry Development.* Appendix 2 of *Cross River National Park, Okwangwo Division: Plan for Developing the Park and Its Support Zone.* World Wide Fund for Nature (WWF, Godalming, UK).

Okafor, J. C. & Caldecott, J. O. (1990)   Using biodiversity and agroforestry systems for conservation in Nigeria. *Proceedings of the International Conference on Conservation of Tropical Biodiversity (Kuala Lumpur, Malaysia), June 12th–16th 1990.*

Okafor, J. C. & Fernandes, E. C. M. (1987)   Compound farms of south-eastern Nigeria: a predominant agroforestry home garden system with crops and small livestock. *Agroforestry systems,* **5**: 153–68.

Okali, D. U. U. (1989)   *Forestry Studies in Conjunction with the Soil Survey and Land Evaluation of the Oban Sector of the Proposed Cross River National Park: Final Report.* Appendix 3 of *Cross River National Park, Oban Division: Plan for Developing the Park and Its Support Zone.* World Wide Fund for Nature (WWF, Godalming, UK).

Okali, D. U. U. (1990)   *Forestry Aspects.* Appendix 3 of *Cross River National Park, Okwangwo Division: Plan for Developing the Park and Its Support Zone.* World Wide Fund for Nature (WWF, Godalming, UK).

Olney, P. J., Mace, G., & Feistner, A. (editors, 1993)   *Creative Conservation: Interactive Management of Wild and Captive Animals.* Chapman & Hall (London, UK).

PADI (1990)   *Open Water Diver Manual.* Professional Association of Underwater Diving Instructors (PADI, Santa Ana, California, USA).

PADI (1994)   *Ten Ways a Diver Can Protect the Underwater Environment.* Professional Association of Diving Instructors (PADI, Santa Ana, California, USA).

Pancho, J. V. (1983)   Vascular Flora of Mt Makiling and Vicinity (Luzon, Philippines), Part I: Kalikasan. *The Philippines Journal of Botany, Supplement No. 1*: 1–15.

Pauly, D. & Christensen, V. (1995)   Primary production required to sustain global fisheries. *Nature*, **374**: 255–7.

PaVo (undated, a)   *Environmental Impacts of Freeport's Grasberg Copper Mine in West Papua*. PaVo Paper No. 1. Papuan People's Study and Information Centre (Delft, the Netherlands).

PaVo (undated, b)   *Freeport's Attack on the Futures of the Papuan Peoples: "Thrusting a spear of economic development into the heartland of West Papua"*. PaVo Paper No. 2. Papuan People's Study and Information Centre (Delft, the Netherlands).

Pearce, F. (1991)   *Green Warriors: the People and the Politics Behind the Environmental Revolution*. The Bodley Head (London, UK).

Peluso, N. L. (1993)   Coercing conservation? The politics of state resource control. *Global Environmental Change*, June 1993: 199–217.

Peterson, M. N. A. (1992)   Diversity of life: its meaning and significance in oceanic realms. Pp. 1–22 in *Diversity of Oceanic Life: an Evaluative Review* (edited by M. N. A. Peterson). Center for Strategic and International Studies (Washington, DC, USA).

Petocz, R. G. (1983)   *Recommended Reserves for Irian Jaya Province*. World Conservation Union (IUCN) and World Wide Fund for Nature (WWF, Jayapura and Bogor, Indonesia).

Petocz, R. G. (1984)   *Conservation and Development in Irian Jaya: a Strategy for Rational Resource Utilization*. Directorate General of Forest Protection and Nature Conservation (Bogor, Indonesia).

Petocz, R. G. (1988)   *Philippines Strategy for Environmental Conservation*. World Wide Fund for Nature (WWF, Washington, DC, USA).

Petocz, R. G. (1989)   *Conservation and Development in Irian Jaya*. E. J. Brill (Leiden, the Netherlands).

Petocz, R. G. & de Fretes, Y. (1983)   *Mammals of the Reserves in Irian Jaya*. WWF/IUCN Project 1528 Special Report, WWF/IUCN Conservation for Development Programme in Indonesia. World Wide Fund for Nature (WWF, Jayapura, Indonesia).

Petocz, R. G., Kirenius, M., & de Fretes, Y. (1983)   *Avifauna of the Reserves in Irian Jaya*. WWF/IUCN Project 1528 Special Report, WWF/IUCN Conservation for Development Programme in Indonesia. World Wide Fund for Nature (WWF, Jayapura, Indonesia).

Petocz, R. G. & Raspado, G. P. (1994)   *Mamalia Darat Irian Jaya*. Gramedia Pustaka Utama (Jakarta, Indonesia).

Pfeffer, P. (1959)   Biologie et migrations du sanglier de Borneo (*Sus barbatus* Muller 1869). *Mammalia*, **23**: 277–303.

Pfeffer, P. & Caldecott, J. O. (1986)   The bearded pig (*Sus barbatus*) in East Kalimantan and Sarawak. *Journal of the Malaysian Branch of the Royal Asiatic Society*, **59**: 81–100.

PHPA (1992)   *Wasur National Park Management Plan*. Directorate General of Forest Protection and Nature Conservation (PHPA) and World Wide Fund for Nature (WWF, Jakarta, Indonesia).

Pimm, S. (1995)   Seeds of our own destruction. *New Scientist*, 8 April 1995 (1972): 31–5.

Pleumarom, A. (1994)   The political economy of tourism. *The Ecologist*, **24**(4): 142–8.

Plotkin, M. & Famolare, L. (editors, 1992) *Sustainable Harvest and Marketing of Rain Forest Products.* Island Press (Washington, DC, USA).

Prasodjo, S. (1992) Evaluation of forest concession maps: a case study using GPS and GIS. Pp. 65–78 in *Voices From the Field: Fifth Annual Social Forestry Writing Workshop* (edited by J. Fox, A. Flavelle, & N. Podger). East–West Centre (Honolulu, Hawai'i, USA).

Pratt, B. & Boyden, J. (1985) *The Field Director's Handbook: an OXFAM Manual for Development Workers.* Oxford University Press (Oxford, UK).

Prescott-Allen, R. & Prescott-Allen, C. (1982) *What's Wildlife Worth?* Earthscan (London, UK).

Primack, R. B. (1991) Logging, conservation and native rights in Sarawak forests from different viewpoints. *Borneo Research Bulletin*, **23**: 3–13.

Primack, R. B. (1993) *Essentials of Conservation Biology.* Sinauer (Sunderland, Massachusetts, USA).

Purseglove, J. W. (1968) *Tropical Crops: Dicotyledons.* Longman (Harlow, UK).

Purseglove, J. W. (1972) *Tropical Crops: Monocotyledons.* Longman (Harlow, UK).

Pye-Smith, C. & Feyerabend, G. B. (1994) *The Wealth of Communities: Stories of Success in Local Environmental Management.* Earthscan (London, UK).

Racuyal, J. T. (1994) *Seasonal Distribution, Relative Abundance and Some Aspects of the Biology of Banana Prawn (*Penaeus merguiensis *de Man) in Southeastern Samar Sea.* Master's Dissertation, University of the Philippines in the Visayas (Iloilo, Panay, Philippines).

Raup, D. M. (1988) Diversity crises in the geological past. Pp. 51–7 in *Biodiversity* (edited by E. O. Wilson). National Academy (Washington, DC, USA).

Ray, G. C. (1988) Ecological diversity in coastal zones and oceans. Pp. 36–50 in *Biodiversity* (edited by E. O. Wilson). National Academy (Washington, DC, USA).

Ray, G. C. (1991) Coastal-zone biodiversity patterns. *BioScience*, **41**(7): 490–8.

Redford, K. H. (1989) Monte Pascoal – indigenous rights and conservation in conflict *Oryx*, **23**(1): 33–6.

Redford, K. H. (1991) The ecologically noble savage. *Cultural Survival Quarterly*, **15**(1): 46–8.

Reid, J. C. (1989) Floral and Faunal Richness of the Oban Division of CRNP. Appendix 7 of *Cross River National Park, Oban Division: Plan for Developing the Park and Its Support Zone.* World Wide Fund for Nature (WWF, Godalming, UK).

Repetto, R. (1988) Overview. Pp. 1–41 in *Public Policies and the Misuse of Forest Resources* (edited by R. Repetto & M. Gillis). Cambridge University Press (Cambridge, UK).

Repetto, R. (1991) *Accounts Overdue: Natural Resource Depletion in Costa Rica.* World Resources Institute (WRI, Washington DC, USA).

Repetto, R., Magrath, W., Wells, M., Beer, C., & Rossini, F. (1989) *Wasting Assets (Natural Resource Accounting in the Economy): II, the Indonesian Resource Accounts.* Word Resources Institute (WRI, Washington DC, USA).

RePPProT (1990) *A National Overview from the Regional Physical Planning Program for Transmigration.* Natural Resources Institute (NRI, Chatham, UK) & Ministry of Transmigration (Jakarta, Indonesia).

Rich, P. V. & Rich, T. H. (1983) The Central American dispersal route: biotic history and paleogeography. Pp. 12–34 in *Costa Rican Natural History* (edited by D. H. Janzen). University of Chicago Press (Chicago, Illinois, USA).

Richardson, S. D. (1990)   Administration, Policy, and Law. Pp. 159–90 in *Forests and Forestry in China*. Environment and Policy Institute, East–West Centre (Honolulu, Hawai'i, USA).

Riney, T. (1982)   *Study and Management of Large Mammals*. Wiley (Chichester, UK).

Roberts, C. M. & Polunin, N. V. C. (1993)   Marine reserves: simple solutions to managing complex fisheries? *Ambio*, **22**(6), September 1993.

Ruddle, K. (1989)   Solving the common property dilemma: village fisheries rights in Japanese coastal waters. In *Common Property Resources: Ecology and Community-Based Sustainable Development* (edited by F. Berkes). World Conservation Union (IUCN, Gland, Switzerland).

Ruddle, K., Hviding, E., & Johannes, R. (1992)   Marine resources management in the context of customary marine tenure. *Marine Resource Economics*, **7**: 249–73.

Ruitenbeek, H. J. (1989)   *Economic Analysis of Issues and Projects Relating to the Establishment of the Proposed Cross River National Park*. Appendix 12 of *Cross River National Park, Oban Division: Plan for Developing the Park and Its Support Zone*. World Wide Fund for Nature (WWF, Godalming, UK).

Ruitenbeek, H. J. (1990a)   *Economic Appraisal*. Appendix 7 of *Cross River National Park, Okwangwo Division: Plan for Developing the Park and Its Support Zone*. World Wide Fund for Nature (WWF, Godalming, UK).

Ruitenbeek, H. J. (1990b)   *Economic Analysis of Conservation Initiatives: Examples from West Africa*. World Wide Fund for Nature (WWF, Godalming, UK).

Russ, G. R. & Alcala, A. C. (1994)   Sumilon Island Reserve: 20 years of hopes and frustrations. *NAGA, The ICLARM Quarterly*, July 1994: 8–12.

Sale, J. B. (1981)   *The Importance and Values of Wild Plants and Animals in Africa*. World Conservation Union (IUCN, Gland, Switzerland).

Sale, P. F. (editor, 1991)   *The Ecology of Fishes on Coral Reefs*. Academic (London, UK).

Samways, M. J. (1994)   *Insect Conservation Biology*. Chapman & Hall (London, UK).

Sandlund, O. T. (1991)   Costa Rica's INBio: towards sustainable use of natural biodiversity. *Norwegian Institute for Nature Research, NINA Notat*, **007**: 1–25.

Sayer, J. A., Harcourt, C. S., & Collins, N. M. (1992)   *The Conservation Atlas of Tropical Forests – Africa*. Macmillan (London, UK).

Schaller, G. B. (1993)   *The Last Panda*. Chicago University Press (Chicago, Illinois, USA).

Schoen, R-J. & Djohani, R. H. (1992)   *A Communication Strategy to Support Marine Conservation Policies, Programmes and Projects in Indonesia for the Years 1992–1995*. World Wide Fund for Nature (WWF, Jakarta, Indonesia).

Schuerholz, G., Heard, S. C., & Sullivan, F. (1989)   *Nigeria: Conservation and Development of the Cross River National Park Section South - Management Plan*. Appendix 5 of *Cross River National Park, Oban Division: Plan for Developing the Park and Its Support Zone*. World Wide Fund for Nature (WWF, Godalming, UK).

Schuerholz, G., Ojong, O., & Bisong, P. (1990)   Okwangwo Division Management Plan. Appendix 5 of *Cross River National Park, Okwangwo Division: Plan for Developing the Park and Its Support Zone*. World Wide Fund for Nature (WWF, Godalming, UK).

Schweithelm, J., Wirawan, N., Elliott, J., & Khan, A. (1992)   *Sulawesi Parks Program: Land Use and Socio-Economic Survey Lore Lindu National Park and Morowali Nature Reserve*. The Nature Conservancy (TNC, Jakarta, Indonesia).

Scott, D. (1989) *A Directory of Asian Wetlands*. World Conservation Union (IUCN, Gland, Switzerland).

Shackleton, Lord (1988) Sarawak natives under threat. *The Times (London)*, 5 January 1988.

Siakor, R. & Ashton-Jones, N. J. (1992) Say no! to endangered species' meat. *Pangolin*, **1**: 5–6 (Calabar, Cross River State, Nigeria).

Silzer, P. J. & Heikkinen, H. (1984) *Index of Irian Jaya Languages*. Summer Institute of Linguistics (Jayapura, Irian Jaya, Indonesia).

Singh, J., Hewawasam, I., McCarney, R., Rose, E., & Ruitenbeek, J. (1990) *Towards the Development of an Environmental Action Plan for Nigeria*. The World Bank (Washington, DC, USA).

Sirait, M., Prasodjo, S., Podger, N., Flavelle, A., & Fox, J. (1994) Mapping customary land in East Kalimantan, Indonesia: a tool for forest management. *Ambio*, **23**: 411–17.

Sittenfeld, A. & Gámez, R. (1993) Biodiversity prospecting by INBio. Pp. 69–97 in *Biodiversity Prospecting* (edited by W. V. Reid, A. Sittenfeld, S. A. Laird, D. H. Janzen, C. A. Meyer, M. A. Gollin, R. Gámez, & C. Juma). World Resources Institute (Washington, DC, USA).

SKEPHI (1990) A Complete Account on Irian. *Setiakawan* (Journal of the NGO Network for Forest Conservation in Indonesia), January–June 1990: 1–95.

SMAP (1992a) *Southern Mindanao Agricultural Programme Profile*. Southern Mindanao Agricultural Programme (Davao City, Philippines) and Commission of the European Communities (Brussels, Belgium).

SMAP (1992b) *SALT Extension Package (Protecting Your Soil for Better Yield): Training Plans*. Southern Mindanao Agricultural Programme (Davao City, Philippines) and Commission of the European Communities (Brussels, Belgium).

SMAP (1993a) *Guidelines on Micro-Project Development and Management*. Southern Mindanao Agricultural Programme (Davao City, Philippines) and Commission of the European Communities (Brussels, Belgium).

SMAP (1993b) *A Manual on SMAP Development Process for Participative Community-Based Rural Development*. Southern Mindanao Agricultural Programme (Davao City, Philippines) and Commission of the European Communities (Brussels, Belgium).

Smil, V. (1983) Deforestation in China. *Ambio*, **12**: 226–31.

Smil, V. (1984) *The Bad Earth: Environmental Degradation in China*. Sharpe (Armonk, New York, USA).

Smil, V. (1992) China's environment in the 1980s: some critical changes. *Ambio*, **21**: 431–6.

Smith, R. A. (1994) Planning and management for coastal eco-tourism in Indonesia: a regional perspective. *The Indonesian Quarterly*, **22**: 148–57.

SPREP (1992) *The South Pacific Biodiversity Conservation Programme: Interim Project Document for the Global Environment Facility*. South Pacific Regional Environment Programme (SPREP, Apia, Western Samoa).

Stark, M. (1992a) *WWF Irian Jaya Conservation Programme, WWF Project 3770: Irian Jaya, Indonesia Programme, Semi-annual Progress Report, July–December 1991*. World Wide Fund for Nature (WWF, Jayapura and Jakarta, Indonesia).

Stark, M. (1992b) *WWF Irian Jaya Conservation Programme, WWF Project 3770: Irian Jaya, Indonesia Programme, Semi-annual Progress Report,*

January–June 1992. World Wide Fund for Nature (WWF, Jayapura and Jakarta, Indonesia).

Stiles, D. (1994)   Tribals and trade: a strategy for cultural and ecological survival. *Ambio,* **23**: 106–11.

Suparlan, P. (1994)   The diversity of cultures in Irian Jaya. *The Indonesian Quarterly,* **22**: 170–82.

Taege, M. & Obiekezie, A. (1989)   *Aquaculture Development in the Support Zone of Proposed Oban National Park.* Appendix 4 of *Cross River National Park, Oban Division: Plan for Developing the Park and Its Support Zone.* World Wide Fund for Nature (WWF, Godalming, UK).

Taege, M. & Obiekezie, A. (1990)   *Aquaculture Development.* Appendix 8 of *Cross River National Park, Okwangwo Division: Plan for Developing the Park and Its Support Zone.* World Wide Fund for Nature (WWF, Godalming, UK).

Tan, J. G. (1993)   Poverty alleviation through an integrated approach to coastal resources management: CERD-Firmed experience in Daram, Samar. Pp 39–49 in *Our Sea Our Life.* College of Social Work and Community Development, University of the Philippines, Siliman (Dumaguete City, Philippines).

Tang, H. T. (1987)   Problems and strategies in regenerating dipterocarp forests in Malaysia. Pp 23–46 in *Natural Management of Tropical Moist Forest: Silvicultural and Management Prospects for Sustained Utilization* (edited by F. Mergen & J. R. Vincent). Yale University (New Haven, Connecticut, USA).

Tangley, L. (1990)   Cataloging Costa Rica's diversity. *BioScience* **40**:633–6.

Tchounkoue, P., Jenkin, R. N., Wicks, C., & Ruitenbeek, H. J. (1990)   *Republic of Cameroon, the Korup Project: Plan for Developing the Korup National Park and its Support Zone.* World Wide Fund for Nature (WWF, Godalming, UK) and Natural Resources Institute (NRI, Chatham, UK).

Tenaza, R. (1990)   The palm oil monster. *BBC Wildlife,* **8**(12): 830–1.

Third, D. & Gibbons, P. (1989)   *Preliminary Building Designs.* Appendix 10 of *Cross River National Park, Oban Division: Plan for Developing the Park and Its Support Zone.* World Wide Fund for Nature (WWF, Godalming, UK).

Thomas, D. W. & Tobias, M. F. (1987)   *Medicinal and Food Plants from Cameroon's Forests: Development and Conservation.* Food and Agriculture Organization of the United Nations (FAO, Rome, Italy) and Centre for the Study of Medicinal Plants (Yaoundé, Cameroon).

Thomas, D. W., Thomas, J.McC., Bromley, W. A., & Mbenkum, F. T. (1989)   *Korup Ethnobotany Survey.* World Wide Fund for Nature (WWF, Godalming, UK).

Tilak, J. B. G. (1989)   *Education and Its Relation to Economic Growth, Poverty, and Income Distribution: Past Evidence and Further Analysis.* World Bank Discussion Papers, 46. The World Bank (Washington, DC, USA).

Tobias, D. & Mendelsohn, R. (1991)   Valuing ecotourism in a tropical rain-forest reserve. *Ambio,* **20**: 91–3.

Tomascik, T., Tomascik, A-M., Moosa, K., & Suharto (in press).   *The Ecology of Indonesian Seas.* Periplus (Singapore).

Turner, R. K., Pearce, D., & Bateman, I. (1994)   *Environmental Economics: an Elementary Introduction.* Harvester Wheatsheaf (Hemel Hempstead, UK).

UNCED (1992)   *Agenda 21.* United Nations Conference on Environment and Development (Rio de Janeiro, Brazil), United Nations Environment Programme (Nairobi, Kenya) and United Nations Secretariat (New York, USA).

UNDP (1991) *Human Development Report 1991.* United Nations Development Programme and Oxford University Press (Oxford, UK).

UNDP (1993) *Human Development Report 1993.* United Nations Development Programme and Oxford University Press (Oxford, UK).

UNEP & IUCN (1988) *Coral Reefs of the World, Volume 2: Indian Ocean, Red Sea and Gulf.* United Nations Environment Programme (UNEP, Nairobi, Kenya) and World Conservation Union (IUCN, Gland, Switzerland).

UNFPA (1991) *The State of the World Population.* United Nations Fund for Population Activities (UNFPA, New York, USA)

Valentine, P. S. (1992) Critical issues in developing ecotourism in the Pacific. Pp. 8–13 in *Ecotourism Business in the Pacific: Promoting a Sustainable Experience* (edited by J. E. Hay). University of Auckland (Auckland, New Zealand) and East–West Center (Honolulu, Hawai'i, USA).

van Lavieren, L. P. (1983) *Wildlife Management in the Tropics with Special Emphasis on South-east Asia: a Guidebook for the Warden.* Ciawi School of Environmental Conservation Management (Bogor, Indonesia).

van Steenis, C. G. G. J. (1950) The delimitation of Malaysia and its main plant geographical divisions. *Flora Malesiana, Series I,* **1**: lxx–lxxv.

Veron, J. E. N. (1993) *Corals of Australia and the Indo-Pacific.* University of Hawai'i Press (Honolulu, Hawai'i, USA).

Wallace, A. R. (1869) *The Malay Archipelago.* MacMillan (London, England).

Ward, J. C. (1989) *Cross River National Parks Telecommunications and Computing Requirements.* Appendix 11 of *Cross River National Park, Oban Division: Plan for Developing the Park and Its Support Zone.* World Wide Fund for Nature (WWF, Godalming, UK).

WCED (1987) *Our Common Future.* World Commission on Environment and Development. Oxford University Press (Oxford, UK).

WCMC (1988) *Nigeria: Conservation of Biological Diversity.* World Conservation Monitoring Centre (WCMC, Cambridge, UK).

WCMC (1992) *Global Biodiversity: Status of the Earth's Living Resources – a Report Compiled by the World Conservation Monitoring Centre* (edited by B. Groombridge). Chapman & Hall (London, UK).

WCMC (1994) *Biodiversity Data Sourcebook* (edited by B. Groombridge). World Conservation Press (Cambridge, UK).

Wells, M. P. (1989) *Can Indonesia's Biological Diversity be Protected by Linking Economic Development with National Park Management? Three Case Studies from the Outer Islands.* The World Bank (Washington, DC, USA).

Wells, M. P. & Brandon, K. E. (1993) The principles and practice of buffer zones and local participation in biodiversity conservation. *Ambio,* **22**: 157–62.

Wells, M., Brandon, K., & Hannah, L. (1992) *People and Parks: an Analysis of Projects Linking Protected Area Management with Local Communities.* The World Bank (Washington, DC, USA).

White, D. (1990) *Okwangwo Division Species Lists.* Appendix 9 of *Cross River National Park, Okwangwo Division: Plan for Developing the Park and Its Support Zone.* World Wide Fund for Nature (WWF, Godalming, UK).

Whitmore, T. C. (1984) *Tropical Rain Forests of the Far East, Second Edition.* Clarendon (Oxford, UK).

Whitmore (1989) Southeast Asian tropical forests. Pp. 195–218 in *Ecosystems of the World, 14B – Tropical Rain Forest Ecosystems: Biogeographical and Ecological Studies* (edited by H. Leith & M. J. A. Werger). Elsevier (Amsterdam).

Whitten, A. J. (1980) *The Kloss Gibbon in Siberut Rain Forest.* Doctoral
   Dissertation, University of Cambridge (Cambridge, UK).
Whitten, A. J. (1982) *The Gibbons of Siberut.* Dent (London, UK).
Whitten, A. J. & Whitten, J. (1992) *Wild Indonesia.* New Holland (London, UK).
Whitten, A. J. & Whitten, J. (1994) *Fishes galore.* Garuda Inflight Magazine
   (Jakarta, Indonesia).
Whitten, A. J., Damanik, S. J., Anwar, J., & Hisyam, N. (1984) *The Ecology of
   Sumatra.* Gadjah Mada University Press (Yogyakarta, Indonesia).
Whitten, A. J., Mustafa, M., & Henderson, G. S. (1987) *The Ecology of
   Sulawesi.* Gadjah Mada University Press (Yogyakarta, Indonesia).
Whitten, A. J., Soeriaatmadja, R. E., & Afiff, S. (in press) *The Ecology of Java
   and Bali.* Periplus (Singapore).
Wilcox, B. A. (1980) Insular ecology and conservation. Pp. 95–117 in
   Conservation Biology: an Evolutionary-Ecological Perspective (edited by M.
   E. Soulé & B. A. Wilcox). Sinauer (Sunderland, Massachusetts, USA)
Wilcox, B. H. R. & Powell, C. B. (editors, 1985) *The Mangrove Ecosystem of
   the Niger Delta: Proceedings of a Workshop.* University of Port Harcourt
   (Port Harcourt, Nigeria).
Wille, C. (1991) Tropical treasures. *Panoscope,* **23**: 15–16.
Williams, M. J. (1994) How science serves fisheries management. *NAGA, The
   ICLARM Quarterly,* July 1994: 13–14.
Winchester, S. (1987) Battle lines in the jungle. *The Guardian (London),* 8
   August 1987.
Worah, S. (1994) *Report on a Trip to Irian Jaya (Jan 23–31).* World Wide Fund
   for Nature (WWF, Godalming, UK).
World Bank (1990) *Indonesia – Sustainable Development of Forests, Lands and
   Waters.* The World Bank (Washington, DC, USA).
World Bank (1991a) *Malaysia – Forestry Subsector Study.* The World Bank
   (Washington, DC, USA).
World Bank (1991b) *China – Environmental Management Study, Volume 1: Main
   Report.* The World Bank (Washington, DC, USA).
World Bank (1993) *Vetiver Grass: the Hedge Against Erosion, Fourth Edition.*
   The World Bank (Washington, DC, USA).
World Bank (1995) *Indonesia – Environmental and Social Aspects of Road
   Network Expansion.* Infrastructure Operations Division, the World Bank
   (Washington, DC, USA).
WRI, IUCN, & UNEP (1992) *Global Biodiversity Strategy: Guidelines for Action
   to Save, Study and Use Earth's Biotic Wealth Sustainably and Equitably.*
   World Resources Institute (WRI, Washington DC), World Conservation Union
   (IUCN, Gland, Switzerland) & United Nations Environment Programme
   (UNEP, Nairobi, Kenya).
WWF (1980) *Saving Siberut: a Conservation Master Plan.* World Wide Fund for
   Nature Indonesia Program (Bogor, Indonesia).
WWF (1985) *Proposals for a Conservation Strategy for Sarawak.* World Wildlife
   Fund Malaysia (Kuala Lumpur, Malaysia).
WWF (1991a) *The WWF Irian Jaya Conservation Programme: Activities Report
   for the ODA, 1990–1991.* World Wide Fund for Nature (WWF, Jayapura and
   Jakarta, Indonesia).
WWF (1991b) *The WWF Irian Jaya Conservation Programme: Progress Report,
   February–April 1991.* World Wide Fund for Nature (WWF, Jayapura and
   Jakarta, Indonesia).

WWF (1991c)   *Sustaining People's Participation in Primary Environmental Care: a Strategy for Conserving Indonesia's Protected Areas through Community-Based Management.* World Wide Fund for Nature (WWF, Jayapura and Jakarta, Indonesia).

WWF (1993a)   *Project Proposal: Creation and Management of the Proposed Lorentz National Park in Irian Jaya, Indonesia.* World Wide Fund for Nature (WWF, Jakarta, Indonesia).

WWF (1993b)   *Development of the Conservation Activities in Irian Jaya, Months of October–December 1992.* World Wide Fund for Nature (WWF, Jayapura and Jakarta, Indonesia).

WWF (1993c)   *Development of the Conservation Activities in Irian Jaya, Months of January–June 1993.* World Wide Fund for Nature (WWF, Jayapura and Jakarta, Indonesia).

WWF (1993d)   *Project Proposal: Integration of Conservation and Development in Nusa Tenggara.* World Wide Fund for Nature (WWF, Jakarta, Indonesia).

WWF (1994a)   *Development of the Conservation Activities in Irian Jaya, Months of July–December 1993.* World Wide Fund for Nature (WWF, Jayapura and Jakarta, Indonesia).

WWF (1994b)   *Development of the Conservation Activities in Irian Jaya, Projects ID0085, ID0103, ID0105, Semester Report Covering January–June 1994.* World Wide Fund for Nature (WWF, Jayapura and Jakarta, Indonesia).

WWF (1994c)   *Strategic Action Plan for the WWF Indonesia Programme.* World Wide Fund for Nature (WWF, Jakarta, Indonesia).

WWF (1994d)   *Sustainable Community-Based Wild Resource Management in Indonesia.* World Wide Fund for Nature (WWF, Jakarta, Indonesia).

WWF (1995a)   *Conservation of Biological Diversity in Irian Jaya.* Draft proposal, 3 March 1995. World Wide Fund for Nature (WWF, Jakarta, Indonesia).

WWF (1995b)   *Limiting the Losses: Interim Measures for Conserving Biological Diversity.* World Wide Fund for Nature (WWF, Godalming, UK).

WWF & IUCN (1989)   *The Botanic Gardens Conservation Strategy.* World Wide Fund for Nature and World Conservation Union (WWF–IUCN, Gland, Switzerland) and IUCN Botanic Gardens Conservation Secretariat (BGCS, Kew, UK).

WWF & IUCN (1995)   *Centres of Plant Diversity: a Guide and Strategy for their Conservation: Volume 1 (Europe, Africa, The Middle East and South West Asia), Volume 2 (Asia and the Pacific) and Volume 3 (The Americas).* World Wide Fund for Nature (WWF, Godalming, UK) & World Conservation Union (IUCN, Gland, Switzerland)

Zerner, C. (1990)   *Common Property Practices and Biological Diversity in the Marine and Coastal Zones of Southeast Asia and the South Pacific.* World Resources Institute (Washington, DC, USA).

Index